# FIELDS OF POWER

The Fight for Farmland inthe Age of Green Energy

By

B.S. Horinga

ISBN

Hardcover: 979-8-90190-613-2

Paperback: 979-8-90190-612-5



BOOK PUBLISHING GROUP LLC

Book Publishing Group LLC

42 Broadway 12th floor, New York, NY 10004

www.bookpublishinggroupllc.com

Dedicated to the farmers who watched the land change around them, yet never stopped fighting for the soil beneath their feet.

# *Table of Contents*

# The Day the Quiet Township Changed

For generations, the land had been predictable. Fields rotated between corn, soybeans, and hay. Families measured time by planting seasons, harvests, and the slow rhythm of rural life. The soil beneath their boots was more than property—it was inheritance, responsibility, and identity.

Then the letters started arriving.

At first, they looked harmless enough: informational packets, invitations to community meetings, glossy brochures filled with images of bright blue solar panels and sleek battery storage facilities. The message was simple and reassuring. Clean energy was coming. Progress was coming. The future was coming.

But as residents began to read more closely, the details became harder to ignore.

Hundreds of acres of farmland would be leased or purchased. Zoning rules would be tested. New industrial-scale infrastructure would appear in places that had always been fields and forests. Decisions about the future of rural communities would move quickly through planning commissions, township boards, and state agencies.

Many people initially assumed the projects would be small or temporary. Others believed local governments would have the final say.

They soon discovered neither assumption was guaranteed.

Across the country, communities like theirs were beginning to face the same question: Who decides what happens to the land?

This book tells the story of that question—and the growing conflict surrounding it.

It is not simply a story about energy. It is about land, governance, rural identity, and the tension between national policy goals and local control. It is about farmers who suddenly found their fields at the center of an energy transition, township officials struggling to interpret complicated regulations, and neighbors discovering that decisions made in distant offices could reshape their communities for decades.

The debate over renewable energy is often framed as a simple choice between progress and resistance. But on the ground, the reality is far more complicated.

For many rural residents, the issue is not whether energy should evolve. The question is how, where, and at what cost.

And in town halls, planning meetings, and kitchen tables across rural America, that debate has only just begun.

# Chapter 1
# The Day the Letters Came

In small rural communities, change usually arrives slowly.

It comes in the form of a new barn being built down the road, a neighbor trying a different crop rotation, or a new family moving into the old farmhouse at the corner of the section line. Life moves with the seasons, and the land anchors everything.

For generations, farmland has carried a quiet certainty. Fields that fed one generation would feed the next. The same fence rows, the same drainage ditches, the same roads running between mile-square sections.

Then, almost overnight, something different began to appear.

The first signs were letters.

They arrived in mailboxes across the countryside—thick envelopes filled with glossy brochures and carefully worded explanations. The pages showed pictures of shining solar panels stretching toward the horizon and modern battery storage facilities promising reliability for the electric grid. The message was optimistic and reassuring. Clean energy was coming, and rural communities had an opportunity to participate.

Some landowners were intrigued. Others were skeptical.

But nearly everyone shared the same assumption: if projects were proposed, the community would have time to discuss them, evaluate them, and decide whether they belonged.

What few realized at the time was that a much larger shift was already underway.

Across the country, energy developers were searching for land—large tracts of land that could host solar arrays, wind turbines, and battery storage facilities. Rural areas, with open fields and relatively low population density, became prime targets.

What began as a few proposals soon turned into something much larger.

Planning commissions began receiving applications for projects covering hundreds of acres. Township officials were asked to interpret ordinances that had never anticipated this kind of development. Residents who had rarely attended public meetings suddenly found themselves sitting in crowded township halls late into the evening.

The conversations grew tense.

Supporters argued that renewable energy represented progress and economic opportunity. Opponents worried about the loss of farmland, the industrialization of rural landscapes, and the long-term consequences for their communities.

Many residents began asking a question that seemed simple on the surface but difficult to answer:

Who decides what happens to the land?

Was the decision local, resting with township governments and the communities they represented?

Or would it ultimately be shaped by state policy, federal incentives, and corporate investment flowing into the renewable energy sector?

As the meetings continued and the proposals multiplied, one thing became clear.

What was unfolding was not simply an energy debate.

It was a struggle over land, authority, and the future of rural America.

# Chapter 2
# The Meeting

The township hall had never been this full.

Pickup trucks lined the gravel parking lot and spilled onto the shoulder of the road. Inside, every chair was taken. People stood along the walls, arms folded, whispering quietly to neighbors they had known for years.

At the front of the room sat the planning commission.

Stacks of documents covered the table—site plans, legal memos, engineering reports, and thick application packets prepared by consultants and attorneys. Words like "megawatts," "grid stability," and "renewable transition" appeared throughout the paperwork.

But for the people in the room, the issue felt much simpler.

Outside those doors were fields.

Fields that had been farmed for generations.

Fields that held memories of planting seasons, harvests, and families who built their lives around the soil beneath their boots.

Now those same fields were being proposed for something entirely different.

Rows of solar panels. Battery storage containers. Security fencing. Industrial infrastructure designed to serve a regional electric grid rather than the local community.

The developers described the project as progress.

Some officials described it as inevitable.

But the people in the room weren't so sure.

One by one, they stepped up to the microphone.

Farmers spoke about the loss of farmland. Residents spoke about safety concerns. Others questioned how decisions of this scale could move forward so quickly through local government processes.

The longer the meeting went, the clearer the tension became.

This was not simply a debate about electricity. It was a debate about land. About local control.

About whether rural communities would have a meaningful voice in decisions that could reshape their landscape for decades.

By the end of the night, no one in that room believed the issue was going away.

And across the country, similar meetings were beginning to unfold.

Quiet townships were discovering they had become the front line of a much larger conflict—one that stretched far beyond their county lines.

# Chapter 3
# The Green Promise

Over the past several decades, renewable energy has been presented to the public as one of the most hopeful and transformative solutions to modern environmental challenges. Solar panels, wind turbines, and advanced battery systems have become powerful symbols of a future where societies can produce the energy they need while dramatically reducing pollution and protecting the planet. Governments, environmental organizations, and energy companies have promoted renewable technologies as part of a broader vision of progress—one in which technological innovation and environmental responsibility move forward together.

This vision, often described as the transition to a "green energy" economy, gained widespread public support through a combination of environmental concern, technological optimism, and political momentum. For many people, renewable energy represents more than just a change in how electricity is produced. It symbolizes a hopeful path toward solving some of the most difficult challenges of the modern world.

## The Environmental Argument

One of the most powerful drivers behind the promotion of renewable energy has been the growing concern about climate change and environmental degradation. Scientists and policymakers have warned that continued reliance on fossil fuels—such as coal, oil, and natural gas—contributes to rising greenhouse gas emissions, which in turn affect global temperatures and weather patterns.

Renewable energy technologies offer an alternative approach. Unlike fossil fuel power plants, solar panels and wind turbines generate electricity without burning fuel, meaning they produce little to no direct emissions during operation. This characteristic has made renewable energy a central component of many climate policy strategies.

Environmental messaging has often focused on the idea that replacing fossil fuels with renewable energy could significantly reduce carbon emissions and slow the pace of climate change. Campaigns promoting renewable energy frequently highlight images of clean landscapes, clear skies, and modern technology working in harmony with nature. These images reinforce the idea that renewable energy represents a cleaner and more sustainable way to power society.

## Technological Optimism

Another reason renewable energy messaging gained such strong traction is the widespread belief in technological progress. Throughout history, technological innovation has repeatedly transformed societies—improving transportation, communication, medicine, and countless other aspects of daily life.

The transition to renewable energy has often been framed within this broader narrative of progress. Just as societies moved from wood to coal, from coal to oil, and from steam engines to modern electricity systems, the shift toward renewable energy is presented as the next stage in an ongoing evolution.

Solar panels and wind turbines are frequently portrayed as cutting-edge technologies that will replace outdated energy systems. Battery storage systems, smart grids, and electric vehicles are often described as part of an integrated energy future powered by innovation. This message resonates strongly in cultures that value technological advancement and problem-solving through engineering and science.

For many supporters, renewable energy represents not only environmental responsibility but also the promise of a modernized energy system built with advanced technologies.

## Economic Opportunity and Job Creation

Economic arguments have also played a major role in building support for renewable energy. Governments and industry leaders often emphasize the potential for renewable energy development to create jobs, attract investment, and stimulate economic growth.

Large solar and wind projects require engineers, construction workers, electricians, and maintenance crews. Manufacturing facilities that produce renewable energy equipment can create additional employment opportunities. In rural areas, developers sometimes promote renewable energy as a new source of economic activity, particularly in regions where traditional industries such as manufacturing or agriculture may face economic pressures.

For landowners, renewable energy projects may also offer financial incentives through land lease agreements. These agreements can provide steady income streams over many years, which can be appealing in agricultural sectors where commodity prices and weather conditions often create financial uncertainty.

By presenting renewable energy as both an environmental and economic solution, policymakers have been able to build support across a wide range of political and social groups.

## Cultural and Political Momentum

Over time, renewable energy has also become a powerful cultural symbol. Solar panels on rooftops, wind turbines on the horizon, and electric vehicles in driveways have come to represent participation in a broader movement toward sustainability.

Political leaders across many countries have incorporated renewable energy into national strategies for addressing climate change and modernizing infrastructure. International agreements and national policies have reinforced the idea that transitioning away from fossil fuels is an important global priority.

As renewable technologies became more visible and widely discussed, they gained momentum not only as policy tools but also as cultural markers of environmental awareness and social responsibility. Supporting renewable energy began to be seen by many as a way to contribute to a collective effort to protect the planet.

## A Compelling Narrative

The success of renewable energy messaging can largely be attributed to how effectively it combined several powerful ideas. It offered a solution to environmental concerns, aligned with long-standing beliefs in technological progress, and promised economic benefits through new industries and jobs. These themes created a compelling narrative that resonated with many people across different political and social backgrounds.

However, as renewable energy projects move from broad policy visions into real-world implementation, communities often encounter complexities that were not always emphasized in the original messaging. Large infrastructure projects, land-use changes, and technical challenges sometimes reveal aspects of the transition that are less visible in promotional narratives.

Understanding how renewable energy was presented to the public—and why that message gained such strong support—is an important step in examining the broader debate about energy policy. The promise of renewable energy remains a powerful and motivating idea, but its implementation raises questions that extend beyond simple slogans.

As societies continue to pursue cleaner energy systems, the conversation is increasingly shifting from the promise of renewable energy to the practical realities of how that promise is carried out on the ground.

# Chapter 4
# The Reality of Green Energy

In recent years, renewable energy has been widely presented as a hopeful and necessary solution to some of the most pressing environmental challenges facing the world today. Solar panels gleaming under bright skies, wind turbines slowly turning on distant ridge lines, and advanced battery storage systems quietly stabilizing the electrical grid have become powerful symbols of a cleaner, more sustainable future. Governments, corporations, and environmental organizations often describe renewable energy as a path toward reducing pollution, combating climate change, and modernizing the way societies generate and use electricity. At first glance, the concept appears straightforward: replace fossil fuels with renewable sources such as solar and wind, store excess energy in batteries, and create a new system that produces power without the environmental costs associated with traditional energy production.

This vision has been widely embraced because it promises multiple benefits at once. Renewable energy is frequently described as "clean" because it does not require burning fuel to generate electricity. It is also framed as a technological advancement that represents progress, similar to past transitions that replaced older systems with newer and more efficient innovations. In addition, renewable projects are often promoted as economic opportunities, bringing investment, tax revenue, and job creation to the regions where they are built. Together, these ideas form a compelling narrative that suggests the transition to renewable energy is not only environmentally responsible but also economically beneficial and technologically inevitable.

However, when renewable energy projects move from national policy discussions and promotional materials into real landscapes and communities, the reality can become far more complex. Large-scale renewable energy infrastructure requires significant land, equipment, and supporting systems to function effectively. Unlike traditional power plants, which concentrate energy production in a single location, renewable energy generation often spreads across large geographic areas. Solar farms may occupy hundreds or even thousands of acres of land. Wind turbines must be spaced across wide landscapes to capture sufficient wind resources. Battery energy storage systems require industrial facilities filled with container-sized units designed to store electricity for later use.

Because of these spatial requirements, many renewable energy projects are built in rural areas where land is more available, and population density is lower. For the communities that host these projects, the transition can feel very different from the streamlined narrative often presented at the national level. Residents may suddenly find that farmland, forests, or open landscapes that have remained largely unchanged for generations are now being proposed as sites for major energy developments.

One of the most striking aspects for many communities encountering renewable projects is the scale involved. A few solar panels on a rooftop or a small community installation may appear modest, but utility-scale renewable facilities operate on a much larger level. A single large solar project can include hundreds of thousands of panels arranged in long rows stretching across vast areas of land. These sites often include security fencing, maintenance roads, electrical substations, and extensive wiring systems. Battery storage facilities can consist of multiple rows of large container units connected to cooling equipment, monitoring systems, and safety infrastructure.

In addition to the projects themselves, the transition to renewable energy also requires significant upgrades to the electrical grid. New transmission lines must often be constructed to carry electricity from rural energy sites to urban areas where demand is highest. Substations, transformers, and grid modernization projects become necessary to manage the flow of power generated by variable sources such as wind and solar. These infrastructure developments can extend far beyond the boundaries of individual renewable energy facilities, altering landscapes and land use patterns across entire regions.

Another factor that shapes renewable energy development is the role of financial incentives. Governments frequently provide tax credits, grants, and other forms of support to encourage investment in renewable energy technologies. These incentives are designed to accelerate the transition away from fossil fuels and help emerging technologies compete in the energy marketplace. While these policies have contributed significantly to the rapid growth of renewable energy, they can also create strong financial motivations for developers to build projects quickly in order to qualify for available incentives. In some cases, this can lead to situations where projects move rapidly through planning processes, leaving communities struggling to fully understand the long-term implications.

As renewable energy proposals reach the local level, communities often respond in a variety of ways. Some residents welcome these projects, viewing them as opportunities for economic growth, additional income through land leases, or contributions to environmental goals. Others express concerns about how such developments might affect farmland, wildlife habitat, rural landscapes, property values, and community identity. Debates over renewable energy can sometimes divide communities that have long shared common values and traditions.

Local planning meetings and township hearings frequently become the arenas where these discussions take place. Residents step forward to voice their perspectives, local officials attempt to interpret zoning regulations and development proposals, and developers present plans designed to meet both regulatory requirements and energy production goals. What may have seemed like a distant policy discussion at the national level becomes a tangible and sometimes contentious issue within the community.

The tension that emerges often reflects a deeper gap between the broad vision of renewable energy and the specific experiences of communities hosting these projects. At the policy level, renewable energy represents a strategy for addressing global environmental challenges and transforming energy systems for the future. At the local level, however, these same projects can feel like major industrial developments that permanently reshape familiar landscapes.

Recognizing this gap does not necessarily mean rejecting the transition to renewable energy. Instead, it highlights the importance of understanding the full scope of what such a transition entails. Energy systems have always required infrastructure, land, and resources. Every major energy shift throughout history—from coal to oil to electricity—has reshaped economies and landscapes in significant ways.

The current transition toward renewable energy is no exception. While it offers potential environmental benefits, it also raises important questions about land use, infrastructure development, community participation, and long-term stewardship of the land. Understanding both the promise and the reality of green energy allows for a more balanced and informed discussion about how societies can pursue environmental goals while also respecting the communities and landscapes that make such transitions possible.

Ultimately, the debate surrounding renewable energy is not simply about technology or electricity. It is about how societies choose to balance global priorities with local experiences, and how communities participate in shaping decisions that affect the land they call home.

# Chapter 5
# The Energy Land Rush

In recent years, a dramatic shift has taken place across many rural regions as energy developers began searching for large tracts of land suitable for renewable energy projects. What started as a gradual expansion of solar and wind installations has, in some areas, accelerated into what many observers describe as an "energy land rush." Driven by government incentives, corporate investment, and ambitious renewable energy targets, developers began looking beyond urban areas and industrial zones to the wide open landscapes of rural America.

For many communities, the arrival of renewable energy developers came suddenly. Landowners began receiving letters, phone calls, and visits offering lease agreements for solar arrays, wind turbines, or battery storage systems. In some cases, these proposals covered hundreds or even thousands of acres at a time. What had once been farmland, pastures, or open countryside was now being evaluated as potential infrastructure for the nation's evolving energy grid.

Understanding how and why this land rush occurred requires examining the economic incentives, technological needs, and geographic realities that shape renewable energy development.

## Why Rural Land Became the Target

Renewable energy projects require specific environmental and geographic conditions to operate efficiently. Solar installations need large areas with unobstructed sunlight throughout the day. Wind turbines require locations with consistent wind patterns and sufficient distance between towers. Battery storage systems are often placed near major transmission lines or substations so they can connect directly to the electrical grid.

Rural landscapes frequently provide these conditions. Agricultural land is often relatively flat, open, and free from large buildings or dense development. Population density is lower than in urban areas, which can reduce potential conflicts over noise, visual impacts, or land availability. Large contiguous parcels of land are also more common in farming regions, making it easier to assemble the acreage required for utility-scale projects.

For developers planning large energy facilities, these characteristics make rural land especially attractive. What appears to be open farmland from an agricultural perspective can look like ideal infrastructure space from an energy development perspective.

## Special Land Use in Zoning Law

A Special Land Use (SLU) is a zoning tool used by local governments to allow certain types of development that may be appropriate in a zoning district but require additional review and conditions before they can be approved. Unlike uses that are permitted "by right," special land uses must go through a public review process, typically involving the planning commission and sometimes the township board. The purpose of this process is to ensure that a proposed project is compatible with surrounding land uses, does not create safety concerns, and meets specific standards outlined in the local zoning ordinance. Public hearings are usually required, giving residents an opportunity to voice support, raise concerns, or request additional conditions before a decision is made.

Under zoning laws in many states, including those guided by the Michigan Zoning Enabling Act, special land uses are intended to balance property rights with community protection. Local governments may approve an SLU only if the applicant demonstrates that the project meets all ordinance standards and will not negatively impact public health, safety, or welfare. Conditions may be placed on the approval to address issues such as setbacks, traffic, environmental protection, or operational limits. Because of this structured review process, special land use decisions often become a central point of debate in communities when large projects—such as industrial facilities, renewable energy installations, or other major developments—are proposed within agricultural or rural zoning districts.

Energy developers and other large infrastructure companies often use the Special Land Use (SLU) process as a pathway to build projects in zoning districts where the use is not automatically permitted but may be allowed with approval. By applying for an SLU, companies can propose projects—such as large solar installations, wind farms, or battery storage facilities—in agricultural or rural zones and argue that the use fits within the conditional framework of the ordinance. Because SLU decisions focus on whether a project meets specific criteria rather than whether the community wants that type of development in the district at all, the process can sometimes shift the debate toward technical compliance rather than broader land-use policy questions. As a result, planning commissions and local officials may find themselves evaluating whether a proposal satisfies ordinance standards, even when residents believe the scale or nature of the project represents a fundamental change in how the land was originally intended to be used.

## The Role of Financial Incentives

Government policies have played a major role in accelerating renewable energy development. Federal tax credits, state renewable portfolio standards, and various financial incentives have made solar and wind projects increasingly attractive investments for energy companies and investors.

These programs are designed to encourage the rapid expansion of renewable energy capacity in order to reduce greenhouse gas emissions and diversify energy sources. However, they also create strong financial motivations for developers to secure land quickly and begin construction within specific timelines to qualify for available incentives.

As a result, companies often compete to identify and secure suitable project locations before competitors do. Land agents may approach multiple landowners within the same area, offering lease payments in exchange for long-term use of their property. These agreements can sometimes provide landowners with stable income over decades, which can be appealing in agricultural markets where commodity prices fluctuate.

This combination of policy incentives and investment capital has fueled rapid growth in renewable energy development, contributing to the sense that a large-scale land acquisition effort is underway in many rural regions.

## The Scale of Modern Energy Projects

Another factor contributing to the land rush is the sheer scale required for modern renewable energy production. Unlike conventional power plants, which concentrate energy generation in a single facility, solar and wind installations spread energy generation across large areas.

A utility-scale solar project may require hundreds of acres to produce electricity at levels comparable to traditional power plants. Wind farms often include dozens or even hundreds of turbines spaced widely across agricultural landscapes. Battery storage systems designed to support grid stability may consist of rows of large container units connected to substations and transmission infrastructure.

Because renewable energy generation is more dispersed than fossil fuel generation, more land is required to produce the same amount of electricity. As demand for renewable power increases, developers must identify ever-larger areas capable of supporting these installations.

## The Arrival of Land Agents

In many communities, the first sign of this energy land rush comes in the form of land acquisition efforts. Energy companies often hire land agents whose job is to identify suitable locations and negotiate lease agreements with property owners.

These agents typically approach farmers and landowners with proposals outlining potential lease payments and project timelines. In some cases, developers attempt to secure options on large numbers of parcels before publicly announcing the full scope of a project. This approach allows companies to assemble the acreage needed to make a project financially viable.

For some landowners, these offers represent an attractive financial opportunity. Lease payments for energy projects can provide a steady income that is not directly affected by weather, crop prices, or livestock markets. For others, the decision can be more complicated, particularly if they value preserving farmland or maintaining the agricultural character of the area.

As different landowners make different choices, communities can find themselves divided between those who support development and those who are concerned about its long-term impacts.

## Infrastructure Beyond the Project Boundary

Renewable energy development also requires extensive supporting infrastructure beyond the boundaries of individual project sites. Transmission lines must connect new energy facilities to regional electrical grids. Substations must convert and distribute power to utilities and consumers. Access roads and maintenance facilities must be built to support long-term operations.

These infrastructure needs often extend across multiple properties and communities, further expanding the footprint of renewable energy development. Transmission corridors, in particular, can span many miles as electricity generated in rural areas is transported to urban population centers.

For communities that host these projects, the combined impact of generation facilities and transmission infrastructure can significantly alter landscapes that have historically been dominated by agriculture or natural habitats.

## A New Competition for Land

The emergence of renewable energy as a major land use has introduced a new form of competition for rural land. In many regions, farmland is now being evaluated not only for its agricultural productivity but also for its potential to generate electricity.

This shift raises important questions about how land should be allocated among competing needs such as food production, energy generation, housing, conservation, and economic development. While renewable energy may offer important environmental benefits, its expansion also highlights the limited availability of land resources.

For rural communities, the energy land rush represents a moment of significant change. Decisions made during this period will shape landscapes, economies, and land-use patterns for decades to come.

## The Beginning of a Larger Debate

As renewable energy development continues to expand, the rush to secure suitable land has sparked broader conversations about planning, regulation, and community involvement. Residents, local governments, and policymakers are increasingly grappling with how to balance national energy goals with local priorities and land stewardship.

What began as a search for suitable project sites has evolved into a deeper debate about who controls the future of rural land and how societies should manage the transition to new energy systems.

The energy land rush is therefore more than a wave of development proposals. It represents the intersection of policy, economics, technology, and land use—an intersection that is reshaping rural landscapes in ways few communities anticipated only a decade ago.

# Chapter 6
# When Farmland Becomes Industrial Land

For generations, farmland has been one of the most valuable and essential resources in rural communities. It represents not only the foundation of local economies but also the basis of food production, environmental stewardship, and family heritage. Fields that stretch across the countryside are often the result of decades—or even centuries—of careful management by farmers who understand the rhythms of soil, weather, and seasons. These landscapes are not simply open spaces; they are working systems that support crops, livestock, and the livelihoods of countless families.

In recent years, however, many rural areas have begun to experience a new kind of pressure on their farmland. Large-scale energy projects—particularly solar arrays, battery storage facilities, and the transmission infrastructure needed to connect them to the electrical grid—are increasingly being proposed on agricultural land. What were once productive fields may suddenly be targeted as sites for industrial-scale energy development.

This transformation raises important questions about how land is used, how rural communities evolve, and how societies balance competing priorities such as energy production and food security.

## Loss of Prime Agricultural Land

One of the most immediate concerns in the conversion of farmland to energy infrastructure is the loss of productive agricultural acreage. Many renewable energy projects require large, relatively flat areas of land with good access to sunlight or transmission lines—conditions that often overlap with prime farmland.

When hundreds or thousands of acres of cropland are converted into solar fields or other energy facilities, that land is effectively removed from agricultural production for the life of the project, which can last twenty-five to thirty years or longer. While some developers argue that the land may eventually be restored after decommissioning, the reality is that the land will no longer function as a working farm during that entire period.

For farming regions that rely heavily on agriculture, this change can have significant ripple effects. Local grain elevators, equipment suppliers, and agricultural service businesses depend on a stable base of farming activity. When farmland disappears, even gradually, these interconnected parts of the rural economy can begin to weaken.

## Soil Impacts and Land Restoration Questions

Another important issue involves the condition of the soil itself. Healthy farmland is not simply dirt; it is a complex biological system that includes organic matter, microorganisms, insects, and plant roots working together to support crop growth.

Construction of large energy projects often involves heavy equipment, grading, and the installation of structural supports, access roads, and electrical infrastructure. These activities can compact soil layers, alter drainage patterns, and disrupt soil biology. Once soil is compacted, it can become more difficult for water to penetrate and for plant roots to grow deeply.

Over time, the absence of traditional crop rotations and soil management practices may also affect soil health. Farmers typically maintain soil fertility through careful use of cover crops, nutrient management, and conservation practices. When land is removed from agriculture, those systems may no longer operate in the same way.

Developers sometimes promise that farmland will be restored once projects are decommissioned decades later. However, questions remain about how easily soil that has been altered by construction and long-term infrastructure can return to its previous productivity. Restoration may require extensive remediation, and the original soil structure may never fully recover.

## Rural Character and Generational Farming

Beyond economics and soil science, farmland also plays a central role in the cultural identity of rural communities. Many farms are passed down through generations, creating a deep sense of continuity and stewardship. Fields are not just parcels of land but part of family histories that stretch back decades.

When farmland is converted into energy infrastructure, it can fundamentally change the visual and cultural landscape of a rural area. Rows of solar panels, security fencing, industrial equipment, and transmission lines can replace open fields and traditional agricultural scenery. For some residents, this shift represents progress and modernization. For others, it feels like the erosion of a way of life that has defined rural communities for generations.

The transformation can also create difficult decisions within farming families themselves. Some landowners may choose to lease their land for energy projects, attracted by the stable income such agreements can provide. Others may wish to continue farming and preserve the agricultural character of the land. These choices can sometimes create tension between neighbors who have long shared the same rural environment but now face different economic pressures and priorities.

## Food Security vs. Energy Policy

At a broader level, the conversion of farmland to energy infrastructure raises questions about the relationship between food production and energy policy. Agricultural land is a finite resource, and societies depend on it to produce the crops and livestock that feed growing populations.

As renewable energy projects expand, policymakers must consider how much farmland can reasonably be converted without affecting long-term food production. Some advocates argue that renewable energy can coexist with agriculture through practices such as dual-use systems, where crops or grazing occur beneath solar panels. Others remain skeptical about whether such approaches can truly replace the productivity of traditional farming.

Balancing these priorities becomes increasingly important as both energy demand and global food needs continue to rise. Energy transitions may be necessary to address environmental challenges, but they must also account for the importance of preserving fertile land capable of sustaining agriculture for future generations.

## A Long-Term Decision

Ultimately, the transformation of farmland into industrial energy sites represents a long-term decision with consequences that extend far beyond individual projects. Land that has supported farming communities for decades may be repurposed for an entirely different function—one that reshapes local economies, landscapes, and cultural identities.

As rural communities confront these changes, they are often forced to grapple with difficult questions about land stewardship, economic opportunity, and the future of agriculture. The decisions made today will influence not only the energy systems of the future but also the landscapes and livelihoods that define rural life.

Understanding these trade-offs is essential for creating policies that respect both the need for energy development and the enduring value of farmland as one of society's most fundamental resources.

# Chapter 7
# The Science of Soil

Healthy farmland is far more than a surface layer of dirt capable of supporting crops. Beneath every productive field lies a complex living system composed of minerals, organic matter, water, air, microorganisms, insects, and plant roots. This system has developed over decades—sometimes centuries—through natural processes and careful agricultural stewardship. Farmers rely on the health of this soil system to grow crops, maintain yields, and preserve land productivity for future generations.

When large infrastructure projects are constructed on farmland, the soil beneath that land can be significantly affected. Even when the surface appears relatively unchanged, activities such as heavy equipment operation, grading, installation of support structures, and long-term changes in land management may alter the delicate balance that allows soil to function effectively.

Understanding soil science helps explain why the conversion of farmland into energy infrastructure raises questions about long-term land health and restoration.

## Soil Compaction

One of the most immediate impacts of large construction projects on farmland is soil compaction. Compaction occurs when heavy machinery presses down on the soil surface, compressing the spaces between soil particles.

Healthy soil contains a network of pores—tiny gaps between particles that allow air and water to move through the ground. These pore spaces are essential for root growth, oxygen exchange, and microbial activity.

When soil becomes compacted, these pores collapse or shrink, making it more difficult for water to penetrate and for roots to expand.

Construction equipment used in large energy projects can be extremely heavy. Trucks, cranes, pile drivers, and earth-moving machinery may repeatedly travel across the same areas during installation. Over time, this repeated pressure can create dense layers within the soil that restrict plant growth.

Compacted soil may also reduce crop yields if the land is returned to agricultural production later. Farmers often use techniques such as deep tillage to break up compacted layers, but restoring natural soil structure can take many years.

## Water Flow and Drainage

Water movement through soil is another critical component of soil health. In productive farmland, rainfall typically infiltrates the soil gradually, replenishing groundwater and providing moisture to crops. Soil structure plays a key role in this process by allowing water to flow through channels created by roots, organisms, and natural soil formation.

When soil becomes compacted or disturbed, water flow patterns can change. Compacted soil may prevent water from penetrating the ground efficiently, causing water to pool on the surface or run off more quickly during heavy rain. This can increase the risk of erosion and reduce the amount of water available to plants.

Construction of infrastructure such as gravel roads, foundations, and drainage systems may also alter how water moves across a landscape. Even relatively small changes in drainage patterns can affect neighboring farmland by shifting water flow toward or away from certain areas.

Maintaining proper drainage is particularly important in agricultural regions where farmers have invested significant effort in managing water through tile drainage systems and other soil conservation practices.

## Microbial Life Beneath the Surface

Perhaps the most remarkable aspect of soil is the vast community of microscopic organisms living within it. Healthy soil contains billions of bacteria, fungi, and other microorganisms that help break down organic matter, cycle nutrients, and support plant growth.

These organisms form complex relationships with plant roots. Certain fungi, known as mycorrhizae, attach to roots and help plants absorb nutrients such as phosphorus from the soil. Bacteria play key roles in converting nitrogen into forms that plants can use. Earthworms and insects help aerate the soil and mix organic matter into deeper layers.

Disturbances to soil structure can affect these biological communities. Compaction reduces oxygen levels in the soil, which may limit microbial activity. Long-term changes in vegetation—such as the removal of crops and their root systems—can alter the availability of organic material that microorganisms depend on for food.

Because soil ecosystems are interconnected, changes in microbial populations can influence the broader health and productivity of the land.

## Vegetation and Soil Cover

Agricultural soils are usually managed through crop rotations, cover crops, and other practices designed to maintain organic matter and prevent erosion. Plants play an essential role in protecting soil by holding it in place with roots and providing organic material as leaves and stems decompose.

When land is converted to energy infrastructure, the vegetation covering the soil often changes. Some renewable energy projects maintain grasses or low-growing plants beneath solar panels to control erosion and support pollinator habitats. While this vegetation can provide some benefits, it does not always replicate the nutrient cycling or soil management practices associated with active farming.

The absence of traditional crop production also means that farmers are no longer applying soil amendments such as compost or manure that help replenish nutrients and maintain organic matter levels.

## Long-Term Restoration Questions

One of the most frequently discussed questions surrounding large energy projects on farmland is whether the land can be fully restored once the infrastructure reaches the end of its operational life. Solar and battery facilities are typically designed to operate for twenty-five to thirty years before major equipment replacement or decommissioning occurs.

At that point, developers may remove panels, support structures, electrical equipment, and other components. However, the condition of the soil beneath these installations may depend on how the land was managed during the life of the project.

If compaction, drainage changes, or loss of soil biology have occurred, restoring the land to its previous agricultural productivity may require significant effort. Soil remediation techniques such as deep tillage, organic amendments, and reintroduction of crop rotations can help rebuild soil structure and microbial activity, but these processes often take time.

In some cases, soil conditions may gradually recover through natural processes once farming resumes. In others, restoration may require careful planning and investment to ensure that the land can once again support productive agriculture.

## Soil as a Living Resource

The science of soil reminds us that farmland is not simply a surface on which activities take place. It is a living resource that supports both agriculture and broader ecological systems. Maintaining soil health requires long-term stewardship and an understanding of how different land uses affect the physical, chemical, and biological properties of the ground.

As energy infrastructure continues to expand into rural landscapes, the condition of the soil beneath these projects becomes an important consideration. The choices made during construction, operation, and eventual decommissioning will influence whether the land can continue to serve future generations as productive farmland.

Understanding soil science, therefore, adds an important layer to the broader discussion about land use and energy development. Beneath every solar panel or wind turbine lies a complex ecosystem whose health will ultimately shape the long-term legacy of these projects on the land.

# Chapter 8
# The Science of Water

## Hydrology and the Possible Consequences of Large Energy Developments

Water has always shaped the landscape of rural communities. Long before roads, buildings, or electrical infrastructure appeared, the movement of water determined where farms could flourish, how soil formed, and how ecosystems developed. Farmers learn to understand their land largely through water—how rain moves across fields, where drainage collects, and how soil holds or releases moisture.

When large infrastructure projects are built on farmland, these natural water patterns can sometimes change. Solar arrays, battery storage facilities, roads, and other construction elements alter the surface of the land and may affect how water moves through soil and across fields. Understanding these potential consequences requires examining the relationship between land development and the natural hydrology of agricultural landscapes.

## Water in Agricultural Systems

Healthy farmland depends on a delicate balance of water. Crops require adequate moisture to grow, but too much water can damage roots and reduce yields. Farmers, therefore, spend considerable effort managing water through drainage systems, soil management practices, and conservation techniques.

In many agricultural regions, drainage tiles—underground pipes installed beneath fields—help remove excess water from the soil and direct it toward ditches, streams, or natural waterways. These systems prevent waterlogging while allowing crops to access the moisture they need during the growing season.

The effectiveness of these drainage systems depends on the natural structure of the soil and the ability of water to move through it. When land is altered through construction or long-term infrastructure placement, these systems can sometimes be affected.

## Surface Runoff and Stormwater

One of the most common concerns related to large energy developments is the way they influence stormwater runoff. When rain falls on open farmland, much of the water gradually infiltrates into the soil, where it replenishes groundwater and supports plant growth.

Construction activities, however, can introduce surfaces that change how water flows across the land. Gravel roads, compacted soil, equipment pads, and drainage systems may cause rainwater to move more quickly across the surface rather than soaking into the ground.

Increased runoff can lead to several potential effects:

- Faster movement of water into nearby ditches and streams
- Increased erosion during heavy rainfall
- Sediment entering waterways
- Altered drainage patterns affecting neighboring fields

For this reason, many energy projects must include stormwater management plans designed to control runoff and prevent flooding or erosion.

## Soil Compaction and Water Infiltration

Another factor influencing water movement is soil compaction caused by heavy construction equipment. When soil becomes compacted, the pore spaces that normally allow water to move through the ground are reduced.

Compacted soil may prevent water from infiltrating effectively, causing rainfall to accumulate on the surface. This can lead to pooling water, increased runoff, or localized flooding during heavy storms.

In agricultural systems, soil structure plays a crucial role in water management. Healthy soil allows rainwater to slowly filter downward, replenishing moisture for crops and reducing erosion. Changes to soil structure can therefore have long-term effects on how water behaves within a landscape.

## Drainage Systems and Infrastructure

Many farmland areas rely on carefully designed drainage networks that include ditches, culverts, and subsurface tile systems. These systems are often interconnected across multiple properties and can extend across entire watersheds.

When new infrastructure is installed, these drainage systems must be carefully mapped and preserved. Construction that disrupts tile lines or alters drainage channels can affect water flow across large areas of farmland.

Developers and engineers often conduct drainage studies before construction begins to ensure that existing systems remain functional. In some cases, tile lines must be relocated or protected to prevent long-term impacts on surrounding fields.

## Wetlands and Natural Waterways

Rural landscapes frequently contain wetlands, streams, and other natural water features that support wildlife and help regulate water flow. Wetlands in particular serve as natural sponges that absorb excess rainfall and slowly release water over time.

Energy development projects located near wetlands or waterways may require environmental review to ensure these ecosystems are protected. Disturbing wetlands can affect water quality, wildlife habitat, and flood control systems.

Environmental regulations often require buffer zones around wetlands and waterways to prevent construction activities from disrupting these sensitive areas.

## Water Quality Considerations

Beyond water movement, some residents express concerns about potential impacts on water quality. Construction activity can introduce sediment into nearby streams or drainage ditches if soil is not properly stabilized.

Stormwater management systems are typically designed to reduce this risk by capturing sediment before it reaches waterways. Vegetative buffers, retention basins, and erosion control measures may be used to protect water quality during and after construction.

In battery storage facilities or substations, additional safeguards are often included to prevent leaks or chemical releases that could affect nearby water sources.

In many rural townships, drinking water does not come from large municipal treatment plants or surface reservoirs. Instead, homes, farms, and local businesses rely on private or small community wells that draw water directly from underground aquifers. An aquifer is a natural layer of permeable rock, sand, or gravel that stores groundwater beneath the surface. Rain and snow gradually filter through soil and rock over time, replenishing this underground supply. Because aquifers recharge slowly and depend on natural infiltration from the land above them, the health of the land surface is closely connected to the long-term reliability of well water.

For communities that depend on wells, protecting the aquifer becomes a critical part of land-use planning. Activities that alter soil structure, increase runoff, or introduce contaminants can potentially affect how water moves through the ground and how clean that water remains. Rural residents often view groundwater as one of their most valuable shared resources, since it supplies drinking water for homes, livestock, and agriculture. As new infrastructure or development is proposed, many townships carefully consider how projects might affect groundwater recharge, water quality, and the long-term sustainability of the aquifer that supports their community.

## Long-Term Land Restoration

Another important question related to water involves the long-term condition of land once energy infrastructure reaches the end of its operational life. Renewable energy facilities may remain in place for several decades, after which they may be removed or replaced.

If the land is returned to agriculture, restoring natural drainage patterns and soil structure may be necessary to ensure water continues to move properly through the soil. Rebuilding healthy soil ecosystems and restoring drainage systems can take time, but may allow farmland to regain its productivity.

Communities often consider these long-term restoration questions when evaluating development proposals.

## Water as a Shared Resource

Water does not recognize property boundaries. Rain falling on one field may eventually flow into neighboring land, streams, or larger watersheds. Because of this interconnected nature, changes in land use can affect water systems beyond the boundaries of a single project site.

Protecting water resources, therefore, requires careful planning, engineering oversight, and ongoing monitoring to ensure that natural hydrological systems continue to function properly.

For rural communities, water remains one of the most valuable resources tied to the land. Farmers depend on it for crops, ecosystems rely on it for habitat, and communities rely on it for drinking water and environmental stability.

As energy infrastructure becomes more common in rural landscapes, understanding the relationship between land development and water systems becomes an important part of ensuring that both agriculture and the environment remain sustainable for generations to come.

# Chapter 9
# The Solar Boom

Over the past decade, solar energy has moved from a niche technology to one of the fastest-growing sources of electricity in the world. Once limited primarily to rooftops and small demonstration projects, solar power is now being deployed at an industrial scale across large areas of land. Governments, utilities, and private investors increasingly view solar energy as a central component of the future energy system.

As a result, large solar developments—often called utility-scale solar farms—are expanding rapidly across rural landscapes. These installations can cover hundreds or even thousands of acres and produce electricity for regional power grids rather than local use.

For many rural communities, the arrival of solar projects marks the beginning of a new chapter in land use. Fields that once produced crops or supported livestock are being evaluated for their ability to generate electricity from sunlight.

Understanding how the solar boom emerged—and how it affects rural landscapes—requires examining both the technology itself and the economic forces driving its expansion.

## The Rise of Utility-Scale Solar

Solar power works through photovoltaic (PV) panels that convert sunlight directly into electricity. Individual solar panels generate relatively small amounts of power, which means large numbers of panels must be installed to produce electricity at a scale meaningful for utilities.

Advances in manufacturing and declining panel costs have made solar energy increasingly competitive with other energy sources. Over the last two decades, the cost of photovoltaic technology has fallen dramatically, making large solar installations financially viable in many regions.

At the same time, governments introduced incentives and renewable energy targets designed to accelerate the transition away from fossil fuels. These policies created strong demand for new solar capacity, prompting developers to search for large areas of land where projects could be built quickly.

Rural farmland soon became one of the most attractive locations for these developments.

## Why Solar Projects Target Farmland

Developers typically look for several key characteristics when selecting sites for large solar farms:

- Large contiguous parcels of land
- Flat or gently sloping terrain
- Minimal tree cover
- Access to transmission lines or substations
- Lower population density

Agricultural regions often meet all of these criteria.

Many farmland areas already contain large open fields that receive unobstructed sunlight throughout the day. The land is often relatively level, making construction easier and less expensive. In addition, rural areas typically have fewer buildings or residential developments that could interfere with large installations.

For developers planning large projects, these conditions make farmland particularly attractive.

However, the same qualities that make farmland ideal for solar development are also what make it valuable for agriculture.

## The Footprint of Solar Farms

The scale of utility-scale solar projects can be difficult to visualize until construction begins.

A typical solar installation requires five to ten acres of land per megawatt of electricity capacity. This means that a project producing 100 megawatts of power may occupy between 500 and 1,000 acres of land.

Within this space, thousands of panels are arranged in long rows mounted on metal racking systems anchored to the ground. Maintenance roads divide sections of the installation, while electrical cables connect panels to inverter stations and substations.

Perimeter fencing surrounds the entire site to protect equipment and restrict access.

From above, these installations resemble vast geometric patterns stretching across the landscape.

For communities accustomed to open farmland, the transformation can be dramatic.

## Solar Panels and Agricultural Land

One of the central debates surrounding large solar developments involves the relationship between solar energy production and agricultural land use.

Supporters argue that solar installations provide a productive use for land while contributing to cleaner energy generation. Some developers promote the idea that solar farms can coexist with certain types of vegetation or grazing beneath the panels.

Others worry that long-term solar projects remove valuable farmland from agricultural production for decades. In regions where agriculture forms the backbone of the local economy, the conversion of farmland to energy infrastructure can raise concerns about the future of food production and rural economies.

Because solar facilities often operate for 25 to 35 years, the decision to install them represents a long-term change in how land is used.

## Construction and Landscape Transformation

Building a solar farm involves far more than placing panels in a field.

Construction typically includes:

- Driving steel posts into the soil to support panel racks
- installing miles of electrical wiring
- constructing gravel access roads
- building inverter stations and electrical substations
- installing security fencing around the perimeter

Heavy equipment used during construction can compact soil and alter drainage patterns. Once complete, the land beneath the panels functions differently from traditional farmland.

While vegetation may remain beneath the panels, the landscape becomes structured around the infrastructure of electricity generation rather than agricultural cycles.

## Economic Incentives and Land Leases

Solar development has also introduced new financial dynamics into rural communities. Developers frequently offer long-term lease agreements to landowners willing to host solar installations.

These leases can provide stable annual income for landowners, sometimes exceeding what the land might generate through crop production alone. For farmers facing retirement with no one to take over the family farm or uncertain commodity prices, such agreements may appear financially attractive.

However, not all landowners make the same choice. Some prefer to continue farming or worry about the long-term consequences of removing farmland from production.

These differing perspectives can sometimes divide communities as neighbors weigh economic opportunities against concerns about land preservation.

## The Solar Debate

The rapid expansion of solar energy has sparked an ongoing debate about the role of farmland in the future energy system. Supporters see solar power as an essential tool for reducing emissions and modernizing electricity generation. Critics question whether large solar farms should occupy productive agricultural land when other locations—such as rooftops, industrial sites, brown-fields or previously developed areas—might also be available.

This debate reflects a broader challenge facing many communities: balancing the need for cleaner energy with the preservation of landscapes that support agriculture and rural livelihoods.

As solar technology continues to expand, these questions are likely to become increasingly important. Decisions about where solar projects are built will shape not only the energy system of the future but also the character of rural land for decades to come.

# Chapter 10
# Battery Storage: The Megapack Era

As solar and wind energy expanded across the electrical grid, a new challenge quickly emerged. Renewable energy sources such as sunlight and wind are not constant. Solar panels produce electricity only when the sun is shining, and wind turbines generate power only when the wind is blowing. Yet modern societies require electricity every hour of every day.

To solve this problem, energy planners increasingly turned to a new technology: grid-scale battery storage systems.

Large battery installations—often referred to as Battery Energy Storage Systems (BESS)—are designed to store electricity when production exceeds demand and release that energy later when the grid needs it. In theory, this technology allows renewable energy to function more reliably by smoothing out fluctuations in generation.

However, the rapid expansion of battery storage has introduced new questions about safety, infrastructure, and long-term management.

## What Grid Batteries Actually Are

Grid-scale battery systems typically consist of large container units that resemble shipping containers. Each container houses thousands of lithium-ion battery cells along with cooling systems, monitoring electronics, and fire suppression equipment.

These containers are arranged in rows across industrial sites and connected to electrical substations that allow the stored energy to be released back into the power grid.

Some installations contain dozens or even hundreds of these battery units, forming facilities capable of storing hundreds of megawatt-hours of electricity.

## The Role of Batteries in Renewable Energy

Battery storage plays several important roles in modern power systems:

- Storing excess solar power produced during sunny periods
- Releasing electricity during evening hours when solar production drops
- Stabilizing grid frequency during fluctuations in supply or demand
- Providing backup power during short-term outages
- Without some form of storage, large amounts of renewable energy could be wasted when production exceeds demand.

## Safety and Thermal Runaway

Despite their benefits, lithium-ion batteries also present unique safety challenges. Under certain conditions, battery cells can overheat and trigger a chain reaction known as thermal runaway. When this occurs, one battery cell overheats and ignites neighboring cells, potentially causing fires that are difficult to extinguish.

Emergency responders have been developing new procedures to address these risks. Modern battery systems include monitoring systems designed to detect early warning signs of overheating. Facilities also include spacing between battery containers and fire suppression equipment to limit the spread of fires.

Even with these safeguards, battery storage remains one of the most technically complex elements of the renewable energy transition.

## The New Battery on the Block: Lithium-Iron Phosphate

In recent years, lithium-iron phosphate (LFP) batteries have emerged as one of the leading technologies for large battery energy storage systems used alongside solar and wind power. Unlike some earlier lithium-ion batteries that rely on metals such as nickel or cobalt, LFP batteries use iron and phosphate in their chemical structure, making them generally more stable and longer-lasting under repeated charging cycles. This stability has made LFP attractive for grid-scale storage projects, where batteries may charge and discharge daily to balance electricity supply and demand. Although these systems are often promoted for their durability and improved thermal stability compared with some other lithium chemistries, they still require careful design, monitoring, and safety systems because they store large amounts of energy in concentrated installations. As renewable energy continues to expand, lithium-iron phosphate batteries are increasingly becoming a central technology supporting modern electrical grids.

## BESS Construction

Battery Energy Storage System (BESS) construction typically begins with extensive site preparation similar to other types of industrial infrastructure. Developers first clear and grade the land, ensuring the site has a stable foundation capable of supporting heavy equipment and electrical infrastructure. Gravel access roads are often installed to allow construction vehicles and maintenance crews to reach different parts of the facility. Concrete pads or reinforced foundations may be poured where battery containers, inverters, and electrical control equipment will be placed. Underground electrical conduits and grounding systems are also installed early in the process to safely manage the flow of electricity and protect the system from electrical faults or lightning strikes.

Once the groundwork is complete, the battery components themselves are delivered and assembled on-site. Modern BESS facilities commonly use containerized battery units—large metal enclosures similar in size to shipping containers—that house rows of lithium battery modules along with cooling, monitoring, and fire protection systems. These containers are arranged in rows and connected through power conversion equipment that allows stored energy to move between the batteries and the electrical grid. The entire facility is typically surrounded by security fencing, safety barriers, and monitoring systems, and it is connected to nearby substations through transformers and high-voltage electrical lines. Although the visible footprint of a battery storage site may appear relatively compact compared with other infrastructure projects, the construction process involves significant electrical engineering and safety planning to ensure reliable long-term operation.

Construction traffic associated with large infrastructure projects can create significant challenges for farmers, particularly during planting and harvest seasons when rural roads are heavily used for moving equipment and crops. Frequent travel by heavy trucks and construction vehicles can damage narrow country roads, create dust and congestion, and make it difficult for large farm machinery to safely access fields and transport harvests.

# Chapter 11
# Wind Turbines -Towers on the Horizon

Long before solar farms and battery storage facilities became common topics in rural planning meetings, wind turbines were among the first large renewable energy structures to appear across agricultural landscapes. Rising hundreds of feet above fields and pastures, modern wind turbines have become some of the most visible symbols of the renewable energy transition. For supporters, they represent innovation and cleaner energy production. For others, they represent a dramatic change in the rural skyline.

Understanding wind turbines requires looking beyond the spinning blades seen on the horizon. These machines are complex systems of engineering, infrastructure, and land use that interact with the surrounding landscape in ways that extend far beyond their physical footprint.

## How Wind Turbines Work

Wind turbines generate electricity by capturing the kinetic energy of moving air. When wind passes across the turbine's blades, it causes them to rotate around a central hub. This rotation turns a shaft connected to a generator inside the turbine's nacelle—the large housing located at the top of the tower.

Inside the nacelle, mechanical energy from the spinning blades is converted into electrical energy. This electricity then travels down cables inside the tower and into underground collection lines that connect multiple turbines to a central substation. From there, the electricity is transmitted into the regional power grid.

Modern wind turbines are significantly larger than earlier designs. Many towers exceed 300 feet in height, with blades that can extend another

200 feet or more. When fully extended, the total height of a turbine can rival that of large urban buildings.

These dimensions allow turbines to reach stronger and more consistent winds at higher elevations, increasing the amount of electricity they can generate.

## The Scale of Wind Farms

Unlike traditional power plants that concentrate electricity generation within a single facility, wind farms distribute turbines across wide areas of land. Because turbines must be spaced apart to prevent airflow disruption, wind farms can stretch across thousands of acres.

Typical turbine spacing ranges from several hundred feet to more than a quarter mile between towers. This spacing ensures that the wake created by one turbine does not interfere with the wind reaching another.

Although each turbine requires only a relatively small foundation footprint, the overall project area can cover large portions of rural landscapes. Access roads must connect each turbine to allow maintenance crews to reach them, and underground cables link the turbines together across the site.

In many cases, agricultural activities such as crop production or grazing can continue around turbines. However, the presence of tall towers and rotating blades introduces a new visual element into landscapes that were once dominated by fields and barns.

## Construction and Foundations

Building a wind turbine involves a substantial construction process. Before the tower itself is erected, crews must excavate large foundations designed to support the immense weight and mechanical forces generated by the turbine.

These foundations often consist of reinforced concrete bases buried deep underground, sometimes measuring dozens of feet in diameter. Steel anchor bolts embedded within the concrete hold the tower securely in place.

Once the foundation is completed, sections of the turbine tower are delivered to the site and assembled using large cranes capable of lifting heavy components hundreds of feet into the air. The nacelle and blades are then mounted on top of the tower.

The construction of access roads and underground electrical lines is also necessary to connect turbines to the project's substation and the broader electrical grid.

## Energy Production and Variability

Wind turbines produce electricity only when wind speeds fall within a certain range. If the wind is too weak, the blades will not rotate fast enough to generate power. If the wind becomes too strong, turbines may shut down temporarily to prevent mechanical damage.

This variability means that wind energy production can fluctuate throughout the day and across seasons. Grid operators must therefore balance wind generation with other energy sources or storage systems to ensure that the electricity supply remains stable.

Despite these fluctuations, wind power has become one of the fastest-growing sources of renewable electricity in many regions.

## Community Perspectives

The arrival of wind turbines in rural areas often generates strong reactions from local communities. Some residents welcome wind development as a source of economic activity and tax revenue. Landowners who host turbines may receive lease payments that provide additional income.

Others express concerns about visual impacts, noise, shadow flicker from rotating blades, and potential effects on wildlife such as birds and bats. Because turbines are visible from long distances, they can significantly alter the appearance of rural landscapes.

These differing perspectives can lead to debates during zoning hearings and planning meetings, where communities attempt to determine how wind energy fits within their vision for the countryside.

## Wind Energy and the Changing Landscape

Wind turbines represent one of the most striking examples of how renewable energy infrastructure is reshaping rural environments. Their towering structures symbolize both technological progress and the complex trade-offs involved in transitioning to new energy systems.

For some communities, wind turbines have become accepted features of the landscape. For others, they remain controversial symbols of industrial development in agricultural regions.

Regardless of perspective, wind turbines illustrate a broader reality of the modern energy transition: generating electricity on a large scale requires physical space and visible infrastructure.

## A New Presence On The Horizon

As wind energy continues to expand, turbines will likely remain prominent features of many rural landscapes. Advances in turbine technology may improve efficiency and reduce the number of turbines needed to generate electricity, but the basic principle—capturing energy from the wind high above the ground—will remain the same.

For rural communities, the sight of turbines on the horizon represents more than a new source of electricity. It represents a changing relationship between land, technology, and the future of energy.

Like the barns, silos, and grain elevators that once symbolized agricultural progress, wind turbines now stand as markers of a new chapter in the evolving story of the countryside.

# Chapter 12
# The Scale of Industrial Renewables

One of the most surprising aspects of modern renewable energy development for many rural communities is the sheer scale of the projects involved. When people first hear about solar panels or wind turbines, they often imagine relatively small installations—perhaps similar to rooftop solar systems or a few turbines on a distant hill. However, utility-scale renewable energy projects designed to power large portions of the electrical grid operate on a much larger scale than most people initially expect.

To produce electricity in amounts comparable to traditional power plants, renewable energy facilities must occupy large areas of land and distribute equipment across wide landscapes. This difference in scale can significantly change how land is used and how rural landscapes appear.

Understanding the true footprint of industrial renewable energy requires comparing it with something familiar: the scale of traditional farming operations.

## Traditional Farms and Agricultural Land

Modern agriculture already operates on large areas of land, especially in regions where row crops such as corn, soybeans, or wheat dominate the landscape. Farms often consist of multiple fields that may span hundreds or even thousands of acres. However, even within these large farms, the land remains part of a continuous agricultural system.

Fields are rotated between crops, planted and harvested each year, and managed using techniques designed to preserve soil health and productivity. Agricultural land also supports a network of related activities, including equipment storage, grain handling, livestock operations, and transportation.

From a visual perspective, farmland tends to maintain a relatively open appearance. Fields change color with the seasons—green during the growing season, golden during harvest, and bare during winter—but the landscape remains largely unobstructed.

Renewable energy projects introduce a very different form of land use, one defined not by seasonal cycles but by permanent infrastructure.

## Solar Farms and Land Requirements

Utility-scale solar installations require extensive acreage in order to generate significant amounts of electricity. Photovoltaic panels convert sunlight into electricity, but each individual panel produces only a modest amount of power. To generate electricity at the scale required for regional power grids, developers must install large numbers of panels arranged across wide areas.

A typical solar facility may require anywhere from five to ten acres of land for each megawatt of electricity produced, depending on panel efficiency, terrain, and project design. This means that a solar farm producing 100 megawatts of electricity could occupy between 500 and 1,000 acres of land.

To visualize this scale, imagine a single project covering an area equivalent to several large farms combined. Within that space, thousands of solar panels are arranged in long rows that stretch across the landscape. Maintenance roads run between sections of panels, while electrical equipment connects the arrays to substations and transmission lines.

From ground level, the repeating rows of panels create a highly structured and uniform environment. From the air, the installation appears as a vast geometric pattern covering the land.

## Wind Turbine Spacing

Wind energy projects operate differently from solar farms, but they also require large areas of land. Wind turbines must be spaced widely apart to prevent turbulence created by one turbine from interfering with the performance of another. As a result, wind farms typically spread turbines across thousands of acres.

Individual turbines may stand hundreds of feet tall, with blades sweeping circles larger than a football field. While the physical footprint of each turbine foundation is relatively small, the spacing required between turbines means that a single wind project may occupy a broad landscape.

In many cases, agricultural activities such as crop production or grazing can continue around wind turbines. However, the visual impact of tall towers and rotating blades can dramatically alter the appearance of rural landscapes that were once defined primarily by fields and barns.

The spacing required for wind energy also means that turbines may be visible across large distances, becoming dominant features on the horizon.

## Comparing Energy Density

One of the key reasons renewable energy projects require large areas of land is related to energy density—the amount of power that can be produced per unit of land. Traditional power plants that burn coal, natural gas, or nuclear fuel concentrate large amounts of energy in relatively small facilities. Because the fuel contains a high amount of stored energy, a single power plant can generate substantial electricity from a limited footprint.

Renewable energy sources, by contrast, rely on natural flows of sunlight and wind that are more diffuse. To capture enough of this energy to supply large populations, renewable technologies must spread across larger areas.

This difference in energy density explains why renewable installations often appear so expensive. The energy they capture is abundant but widely distributed across the environment.

## Visual and Landscape Impacts

For rural residents, the scale of industrial renewable energy can transform familiar landscapes. Fields that once appeared open and continuous may now be divided by rows of panels or dotted with tall wind turbines. Transmission lines and substations may accompany these installations to carry electricity to distant population centers.

Some communities embrace these changes as signs of economic development and participation in the transition to cleaner energy systems. Others view the transformation more cautiously, particularly when projects are located on productive farmland or near residential areas.

The visual scale of renewable infrastructure can become a central part of local debates, as residents weigh the benefits of energy production against the desire to preserve the character of rural landscapes.

## Understanding the True Footprint

Recognizing the scale of industrial renewable energy is essential for understanding the broader energy transition. While solar panels and wind turbines may appear simple when viewed individually, the systems required to generate electricity at a grid scale involve extensive infrastructure and significant land use.

As societies move toward increasing levels of renewable energy production, the question of where these installations will be located—and how they will reshape landscapes—becomes increasingly important. The scale of industrial renewables ensures that the energy transition is not only a technological shift but also a geographic one, altering how land is used across entire regions.

For rural communities, the realization of this scale often marks the moment when the abstract idea of renewable energy becomes a concrete reality on the ground.

# Chapter 13
# Township Democracy Under Pressure

Across rural America, decisions about land use have traditionally been made at the local level. Township boards, planning commissions, and zoning boards of appeals serve as the institutions through which communities guide development and protect the character of the places where they live. These bodies are often composed of local residents—farmers, business owners, and neighbors—who volunteer or serve part-time to help shape how land is used within their township.

For decades, many planning commission meetings were relatively quiet affairs. Discussions might involve small zoning adjustments, residential building permits, or minor changes to local ordinances. Attendance was often limited, and the decisions being made rarely attracted widespread public attention.

However, the rapid expansion of large-scale energy development has placed new pressures on these local institutions. Solar farms covering hundreds of acres, wind turbine projects stretching across miles of countryside, and battery storage facilities tied to regional power grids have introduced complex decisions that many local governments were never originally designed to handle.

As a result, township democracy in many rural communities has entered a new and challenging phase.

## The Role of Planning Commissions

Planning commissions serve as one of the primary tools local governments use to guide development. These commissions review proposed projects, evaluate how they align with local zoning ordinances and master plans, and make recommendations to township boards regarding whether projects should move forward.

The commission's role often involves balancing multiple priorities. Members must consider property rights, economic development opportunities, environmental impacts, and the long-term vision for the community outlined in planning documents.

When large renewable energy projects are proposed, the responsibility placed on planning commissions can become significantly more complex. Applications for solar farms or battery storage facilities may include hundreds or even thousands of pages of technical documentation, engineering reports, and legal analyses. Commissioners—many of whom serve on a volunteer or part-time basis—must review these materials and determine whether projects comply with local regulations.

This process can be difficult even under ideal circumstances, and it becomes even more challenging when projects generate strong public reactions.

## Zoning Battles

Zoning laws form the backbone of local land-use regulation. These rules establish what types of development are allowed in specific areas of a township, such as agricultural, residential, or commercial zones. Zoning helps communities maintain order in development patterns and protect certain types of land uses.

When large energy projects are proposed in agricultural areas, they often raise questions about whether such projects are compatible with existing zoning classifications. Some ordinances allow renewable energy projects through special land use permits, while others may not have been written with modern energy infrastructure in mind.

As a result, zoning hearings surrounding renewable energy projects can become highly contentious. Residents may argue that industrial-scale energy facilities do not belong in agricultural zones designed for farming. Developers may respond that renewable energy serves a public purpose and should be permitted under existing rules.

These disagreements can turn zoning hearings into lengthy debates where legal interpretations, community values, and economic considerations all come into play.

## Pressure on Local Officials

Local officials serving on township boards and planning commissions often find themselves under significant pressure during these debates. Residents who oppose projects may urge officials to protect farmland and rural character, while landowners supporting projects may emphasize property rights and economic opportunities.

Developers and energy companies may also advocate strongly for approval, sometimes bringing legal teams and consultants to present technical information during hearings. The presence of these professional teams can create an imbalance when compared to citizens who are speaking as individuals without similar resources.

At the same time, local officials must ensure that decisions comply with state laws governing zoning and land use. If a township denies a project improperly, it may face legal challenges from developers who argue that their rights under zoning law have been violated.

These pressures place local decision-makers in a difficult position as they attempt to balance community concerns with legal obligations.

## The Question of Local Control

Underlying many of these debates is a broader question about the role of local governments in shaping land-use decisions. Townships historically held significant authority over zoning, allowing communities to determine how land within their boundaries should be developed.

However, as energy infrastructure becomes more closely tied to statewide or national policy goals, some governments have considered shifting decision-making authority away from local officials. State agencies may be granted power to approve certain energy projects, particularly those considered important for meeting energy targets or grid reliability goals.

When this occurs, residents may feel that their ability to influence decisions about their own communities is diminished. The tension between state-level authority and local control has become a central issue in many energy infrastructure debates.

## Citizens Rediscover Local Government

Ironically, the pressure placed on township governance has also revived interest in local democracy. Many residents who previously paid little attention to planning commission meetings now attend regularly.

Community members read zoning ordinances, study development proposals, and participate in public comment periods.

Township halls that once hosted small meetings may suddenly fill with residents eager to express their views. For many communities, these gatherings represent an opportunity to engage directly with the democratic process at its most local level.

While disagreements may be intense, this renewed civic participation reflects a deeper principle: the belief that decisions about land and community development should involve the voices of those who live there.

## A Defining Test for Local Governance

The arrival of large-scale energy development has placed township democracy under pressure in ways few communities anticipated. Planning commissions and township boards must now navigate complex legal frameworks, technical proposals, and deeply felt community concerns.

These debates ultimately test the resilience of local governance. They challenge communities to determine how much authority should remain at the local level and how decisions affecting land use should be made in an era of rapid technological and policy change.

For many rural residents, the township hall has become the place where national energy policy meets local reality. The outcome of these discussions will shape not only individual projects but also the future role of local democracy in guiding how land is used across rural landscapes.

## Legal Battles and Ordinances:

## Moratoriums, Zoning Rewrites, and Lawsuits

As renewable energy projects spread across rural regions, many communities have discovered that the debate over land use does not remain confined to public meetings or planning discussions. Instead, it often moves into the legal arena, where zoning ordinances, state laws, and court decisions become central to determining what projects move forward and under what conditions.

For townships and counties, renewable energy proposals have exposed gaps in ordinances that were often written long before utility-scale solar farms, battery storage facilities, or large wind projects became common. As a result, communities are increasingly turning to legal tools—such as temporary moratoriums, zoning rewrites, and court challenges—to regain control over how development occurs within their boundaries.

These legal strategies have become a defining feature of the rural response to large-scale energy infrastructure.

## Moratoriums: Pressing Pause

One of the most common legal tools used by local governments is the moratorium. A moratorium is a temporary pause on certain types of development while a community reviews and updates its regulations.

When renewable energy proposals begin appearing in large numbers, local officials may realize that existing zoning ordinances do not clearly address these types of projects. Without specific rules governing setbacks, safety standards, or land-use compatibility, planning commissions may struggle to evaluate applications effectively.

A moratorium allows communities to temporarily halt approvals while they study the issue more carefully. During this period, township boards and planning commissions can consult legal experts, review examples from other jurisdictions, and gather input from residents.

Moratoriums are often controversial. Developers sometimes argue that they delay important energy projects or unfairly restrict property rights. Supporters of moratoriums counter that they are necessary to ensure that local governments have adequate time to create clear and enforceable regulations.

In many cases, moratoriums serve as a reset button that allows communities to step back and develop more comprehensive policies.

## Rewriting Zoning Ordinances

Once a community begins examining its existing regulations, the next step often involves rewriting zoning ordinances to address renewable energy infrastructure more directly. These updates may include detailed standards for where energy facilities can be located and how they must operate.

New ordinances may address issues such as:

- Setback requirements between energy facilities and homes, roads, or property lines
- Height restrictions for wind turbines or other structures
- Safety standards for battery storage installations
- Noise limits and visual impact considerations

- Decommissioning requirements for removing infrastructure at the end of a project's life
- Agricultural compatibility rules for projects located on farmland

  These zoning changes aim to provide clarity for both developers and residents. By establishing clear guidelines, communities can ensure that projects are evaluated consistently and that potential impacts are addressed in advance.

However, updating zoning ordinances is not always simple. Renewable energy technology evolves quickly, and regulations must often balance technical complexity with enforceable legal standards.

## Special Land Use Permits and Variances

Another area where legal issues frequently arise involves special land use permits and variances. These mechanisms allow local governments to consider projects that do not perfectly fit within existing zoning rules but may still be appropriate under certain conditions.

Special land use permits typically require developers to meet specific criteria related to safety, environmental protection, and compatibility with surrounding land uses. Planning commissions often review these applications in public hearings where residents can provide input.

Variances, on the other hand, allow property owners to request exceptions from certain zoning requirements. These requests are typically evaluated by zoning boards of appeals and must meet legal standards demonstrating that strict application of the ordinance would create an unnecessary hardship.

Disputes sometimes arise over whether certain energy projects should be considered special uses, variances, or entirely new land-use categories. These legal distinctions can significantly influence whether projects are approved or denied.

## Lawsuits and Court Challenges

When disagreements over zoning decisions become intense, they sometimes lead to legal action. Developers may challenge local governments if they believe zoning regulations have been applied improperly or if project denials violate property rights.

Conversely, residents or community groups may file lawsuits if they believe that approvals were granted without adequate review or in violation of local ordinances.

Court cases involving energy infrastructure often revolve around complex questions of law, including whether local regulations conflict with state policies, whether zoning decisions were made fairly, and whether environmental review requirements were properly followed.

Legal battles can take months or even years to resolve, and their outcomes may set precedents that influence how future projects are handled.

## The Growing Complexity of Land-Use Law

The expansion of renewable energy has added new layers of complexity to land-use law. Local officials must navigate not only their own ordinances but also state statutes, federal policies, and evolving legal interpretations related to energy development.

For communities with limited legal resources, this complexity can create significant challenges. Planning commissioners and township boards often rely on attorneys and consultants to help interpret regulations and guide decision-making.

At the same time, residents who become engaged in these issues frequently find themselves studying zoning codes and legal frameworks in order to participate effectively in the process.

## Law as a Tool of Community Protection

While legal battles can be contentious, they also reflect the functioning of democratic systems designed to manage competing interests. Zoning ordinances, public hearings, and judicial review all serve as mechanisms through which communities attempt to balance development with the protection of local values.

For rural communities facing rapid infrastructure expansion, these legal tools provide a means of asserting control over how land is used. Moratoriums create time for reflection, zoning rewrites establish clearer rules, and court challenges help resolve disputes when interpretations differ.

In many ways, the legal battles surrounding renewable energy development represent a broader struggle to define the boundaries between progress and preservation.

## The Long-Term Impact

As energy systems continue to evolve, the legal frameworks governing land use will likely evolve as well. New technologies, changing policies, and shifting economic conditions will continue to shape how communities regulate development.

The decisions made in courtrooms, township halls, and planning commission meetings today will influence not only individual projects but also the broader relationship between energy infrastructure and rural landscapes.

For many communities, the legal battles underway today represent more than technical disputes over zoning language. They are part of a larger effort to ensure that the rules governing development reflect the interests and values of the people who live on the land.

# Chapter 14
# State vs Local Power

As renewable energy development expands across rural landscapes, an increasingly important question has emerged about who ultimately controls the decisions that shape these projects. While local governments have traditionally managed land-use decisions through zoning ordinances and planning commissions, state agencies and federal policies often play powerful roles in determining how and where energy infrastructure is built.

This interaction between different levels of government creates a complex system in which local communities, state authorities, and national energy strategies must all coexist. In some cases, these levels of authority work together smoothly. In others, they come into conflict, particularly when state or federal goals appear to override local land-use decisions.

Understanding how state and federal policies influence local outcomes is essential for understanding the broader politics of renewable energy development.

## The Traditional Role of Local Government

Historically, land-use decisions in many parts of the United States have been controlled at the local level. Townships, counties, and municipalities adopt zoning ordinances that determine how land within their boundaries may be used. These regulations divide land into categories such as agricultural, residential, commercial, or industrial zones.

Planning commissions and zoning boards are responsible for reviewing development proposals and determining whether they comply with these local rules. Public hearings allow residents to participate in the decision-making process by raising concerns, asking questions, and offering input.

This system of local governance reflects the principle that communities should have a meaningful voice in shaping how their land is developed. Because local officials live within the communities they serve, they are often familiar with the unique environmental, agricultural, and cultural characteristics of the area.

For decades, this approach allowed rural communities to guide development in ways that reflected local priorities.

## State Energy Policies

The expansion of renewable energy has introduced new layers of authority into this system. Many states have adopted policies designed to accelerate the transition toward cleaner energy sources. These policies may include renewable portfolio standards that require utilities to obtain a certain percentage of their electricity from renewable sources.

To meet these goals, state governments often encourage or facilitate the development of large renewable energy projects. In some cases, state agencies may be granted authority to review and approve energy facilities that exceed certain size thresholds.

Supporters of this approach argue that statewide planning helps ensure a reliable energy system and prevents local regulations from blocking projects that are considered important for broader environmental or economic goals.

However, when state agencies take on greater authority over energy siting decisions, local governments may find their traditional control over land use reduced.

## Federal Influence

Federal policy also plays a role in shaping renewable energy development, though often indirectly. National programs that provide tax credits, grants, or loan guarantees can strongly influence where and how projects are built.

For example, federal incentives for renewable energy construction may encourage companies to develop projects quickly in order to qualify for financial benefits. Environmental regulations and permitting requirements can also influence project design and timelines.

In addition, federal agencies oversee certain aspects of energy infrastructure, such as interstate transmission lines, which can cross multiple states and jurisdictions.

Although federal policy rarely dictates specific project locations at the local level, it can create economic conditions that drive development across entire regions.

## Conflicts Between Levels of Government

The intersection of local zoning authority and higher-level energy policy sometimes produces tension. When local governments deny or restrict projects that state agencies view as essential for energy goals, conflicts can arise over which level of government has the final say.

Some states have passed legislation that allows state regulators to override local zoning decisions for large energy projects. In these cases, developers may bypass local governments and seek approval directly from state agencies.

Supporters of such policies argue that they prevent local opposition from slowing the development of critical energy infrastructure. Critics contend that they undermine local democracy by removing decision-making power from the communities most directly affected by these projects.

These conflicts highlight the challenge of balancing statewide policy objectives with the rights of local communities to shape their own landscapes.

## The Perspective of Rural Communities

For many rural residents, the issue is not simply whether renewable energy projects should exist but who should decide where they are located. Communities that host these projects often experience the direct impacts—changes in land use, landscape appearance, and local infrastructure.

When decisions are made primarily at the state or federal levels, some residents feel that their voices carry less weight in determining outcomes that affect their daily lives. Local officials may find themselves attempting to represent community concerns within regulatory frameworks largely designed at higher levels of government.

This dynamic can lead to frustration and a perception that rural communities are being asked to accommodate large infrastructure projects without having meaningful authority over the process.

## Balancing Authority and Responsibility

The debate over state versus local power reflects a broader question about governance in an increasingly interconnected world. Energy systems operate across large geographic regions, and decisions about electricity generation and transmission must consider regional needs.

At the same time, land-use decisions affect specific communities and landscapes. Local governments remain the institutions closest to those impacts.

Finding the right balance between these levels of authority is one of the central challenges of modern energy policy. Too much fragmentation may slow necessary infrastructure development, while too much centralization may weaken the role of local communities in shaping their future.

## A Continuing Debate

As renewable energy continues to expand, the relationship between state and local authority will remain an important part of the conversation. Laws, regulations, and policies will continue to evolve as governments seek ways to manage the transition to new energy systems.

For rural communities, the outcome of this debate will influence not only the development of renewable energy projects but also the broader principle of who has the power to shape the land.

In the end, the question of state versus local power is about more than policy. It is about governance itself—about how societies decide who should make decisions that affect both the landscape and the people who call it home.

# Chapter 15
# Follow the Money

To understand why renewable energy projects are appearing so rapidly across rural landscapes, it is necessary to look beyond the technology itself and examine the financial forces driving their development. Behind every solar installation, wind farm, and battery storage facility lies a complex system of incentives, investment strategies, and government policies designed to accelerate the transition to renewable energy.

For many energy companies and investors, renewable projects represent not only a technological opportunity but also a powerful financial one. Subsidies, tax credits, and market incentives have created conditions that encourage companies to build projects quickly and on a large scale. These financial mechanisms are often the unseen engine behind what many communities experience as a sudden surge of energy development proposals.

Understanding the economics of renewable energy helps explain why companies are racing to secure land, obtain permits, and begin construction across wide areas of the countryside.

## Government Incentives and Tax Credits

One of the most significant drivers of renewable energy development has been government policy designed to encourage cleaner energy production. Many countries, including the United States, have introduced financial incentives to make renewable energy technologies more competitive with traditional fossil fuel power generation.

These incentives often come in the form of tax credits that allow developers to reduce the amount of taxes they owe when they build qualifying renewable energy facilities. For example, solar projects may qualify for investment tax credits based on the cost of installing photovoltaic systems. Wind projects may receive production tax credits tied to the amount of electricity they generate.

By lowering the effective cost of building renewable energy facilities, these incentives help attract private investment into the sector. Developers are able to secure financing more easily, and investors are more willing to fund projects that benefit from predictable financial returns.

However, these programs often operate within specific timeframes. When incentives are scheduled to decrease or expire after certain dates, companies may rush to complete projects before those deadlines in order to qualify for the full financial benefit.

This timing pressure can accelerate development across large areas as companies compete to secure land and approvals before incentives change.

## Investment and Financial Markets

Renewable energy has also become attractive to large financial institutions and investment funds seeking stable long-term returns. Solar farms and wind projects typically operate for decades once constructed, generating predictable revenue through electricity sales.

Utilities and corporate buyers often sign long-term power purchase agreements with renewable energy developers. These agreements guarantee that electricity produced by a project will be purchased at a predetermined price for many years. Such arrangements provide financial stability that investors value.

As a result, renewable energy projects have attracted funding from pension funds, infrastructure investment firms, and major financial institutions. These investors see renewable energy as a growing sector with long-term profitability potential.

This influx of capital has allowed developers to scale up projects quickly, often pursuing multiple installations simultaneously across different regions.

## Land Leasing and Rural Economics

Another important piece of the financial puzzle involves land leases offered to property owners. Developers frequently approach farmers and rural landowners with proposals to lease portions of their land for solar panels, wind turbines, and BESS systems.

These agreements may provide steady annual payments over twenty-five to thirty years. In agricultural markets where crop prices fluctuate and weather conditions create uncertainty, such payments can appear attractive.

For some landowners, leasing land for renewable energy can provide financial stability that traditional farming does not always guarantee. In other cases, families may view lease income as a way to maintain ownership of land that might otherwise be sold.

However, these agreements can also create difficult choices within rural communities. While some landowners benefit financially, others may worry about how large energy installations could affect farmland availability, property values, or the character of the surrounding landscape.

## Corporate Competition

Because government incentives and investment funding are often tied to project timelines, energy companies frequently compete to identify suitable project sites before their competitors do. This competition can contribute to what some observers describe as a rush to secure land.

Developers may send land agents into rural regions to negotiate leases with property owners, assemble large parcels of land, and obtain the necessary permits to begin construction. In some cases, companies pursue multiple projects at once, hoping to build a portfolio of renewable energy assets.

The faster a project can move from proposal to operation, the sooner it can begin generating electricity and financial returns.

## The Broader Economics of the Energy Transition

The financial incentives driving renewable energy development reflect broader policy goals aimed at transforming energy systems. Governments seeking to reduce emissions and diversify energy supplies often use economic tools to encourage investment in new technologies.

At the same time, these incentives raise questions about how energy markets should operate and who ultimately bears the cost of large infrastructure investments. Tax credits and subsidies represent public policy choices that shift some financial risk away from developers and toward taxpayers.

Supporters argue that such investments are necessary to accelerate the transition toward cleaner energy. Critics contend that the financial structure of renewable energy development may encourage rapid expansion without sufficient attention to long-term land-use impacts.

## A Financial Engine Behind the Transition

While renewable energy is often discussed in terms of environmental goals and technological innovation, the pace of development is also shaped by financial forces. Incentives, investment capital, and competitive pressures combine to create strong motivations for companies to build projects quickly.

For rural communities, understanding this financial context can help explain why large energy proposals sometimes appear suddenly and move rapidly through planning processes. The energy transition is not driven solely by environmental ideals; it is also propelled by powerful economic incentives.

Following the money reveals that behind every row of solar panels or tower of wind turbines and fields of battery containers lies a complex financial structure guiding where and how these projects are built.

# Chapter 16
# The Hidden Costs of the Energy Transition

The transition toward renewable energy is often described as both environmentally necessary and economically beneficial. Solar panels, wind turbines, and battery storage systems are commonly presented as clean, modern technologies that can generate electricity without the pollution associated with fossil fuels. Supporters argue that renewable energy offers a path toward reducing greenhouse gas emissions while creating new industries and jobs.

While this vision has attracted widespread support, critics and analysts increasingly argue that the full costs of the energy transition are not always clearly communicated to the public. Large-scale renewable energy systems require significant investments, infrastructure expansion, and resource extraction to operate effectively. When examined more closely, the transition involves not only environmental benefits but also financial, technical, and environmental trade-offs that are often less visible in public discussions.

Understanding these hidden costs does not necessarily diminish the potential value of renewable energy, but it does highlight the complexity of transforming a nation's energy system.

## Subsidies and Taxpayer Funding

One of the most significant financial aspects of renewable energy development involves government incentives. Many renewable technologies receive support through tax credits, grants, loan guarantees, and other financial programs designed to encourage investment and accelerate deployment.

These incentives exist because renewable energy systems often require large upfront capital investments. Solar farms, wind projects, and battery storage facilities involve expensive equipment, land acquisition, engineering, and grid integration. Government programs help reduce financial risk for developers and make projects economically viable.

Examples of such incentives may include production tax credits for wind energy, investment tax credits for solar installations, and various state-level renewable energy mandates that require utilities to purchase a certain percentage of power from renewable sources.

Supporters argue that these incentives are necessary to help emerging technologies compete with long-established fossil fuel industries that have historically received their own forms of government support. Critics, however, raise concerns about the long-term cost to taxpayers and whether subsidies may encourage rapid development without sufficient consideration of local impacts.

In some cases, the availability of incentives can create strong financial motivations for developers to build projects quickly in order to qualify for expiring tax credits or grant programs. This dynamic can sometimes lead to concerns that economic incentives are driving project timelines more than careful long-term planning.

## Grid Reliability and Storage Challenges

Another often overlooked aspect of the renewable energy transition involves the technical challenges associated with maintaining a reliable electrical grid. Solar and wind power are variable energy sources, meaning their output depends on weather conditions and time of day. Solar panels generate electricity when the sun is shining, and wind turbines produce power when wind speeds are sufficient.

Because electricity demand must be balanced with supply at all times, this variability introduces new complexities for grid operators. When renewable generation fluctuates, other energy sources or storage systems must compensate to maintain stability.

Battery energy storage systems have been proposed as one solution to this challenge. Large batteries can store excess electricity generated during periods of high renewable output and release it later when demand increases. However, battery technology still faces limitations related to cost, storage capacity, and lifespan.

In addition to storage, grid operators may need to maintain backup generation sources, expand transmission networks, and implement advanced grid management systems to accommodate increasing levels of renewable energy. These infrastructure upgrades require significant investment and long-term planning.

The overall cost of integrating renewable energy into existing electrical grids, therefore, includes not only the generation facilities themselves but also the supporting systems needed to ensure reliable power delivery.

## Mining and Manufacturing Impacts

Although renewable energy technologies produce electricity without burning fuel, their manufacturing processes depend heavily on raw materials that must be extracted, refined, and transported. Solar panels, wind turbines, and lithium-ion batteries require a range of minerals and metals, including lithium, cobalt, nickel, copper, and rare earth elements.

Mining these materials can involve significant environmental and social impacts. Large-scale mining operations may alter landscapes, generate waste materials, and consume substantial amounts of water and energy. In some parts of the world, mining activities have also raised concerns about labor practices and environmental regulations.

Manufacturing renewable energy equipment also requires industrial processes that consume energy and produce emissions. While the lifetime emissions of renewable energy systems are generally lower than those associated with fossil fuel power plants, the full lifecycle of these technologies—including mining, manufacturing, transportation, and installation—contributes to their overall environmental footprint.

These upstream impacts are often located far from the communities where renewable energy facilities are built, making them less visible to the public. However, they remain an important part of the broader conversation about the sustainability of energy technologies.

## Decommissioning and Long-Term Liability

Another area that receives increasing attention is the long-term responsibility for renewable energy infrastructure once it reaches the end of its operational life. Solar panels, wind turbines, and battery storage systems are designed to operate for several decades, but eventually they must be removed, recycled, or replaced.

Decommissioning large renewable energy projects can involve significant costs. Solar installations may require the removal of hundreds of thousands of panels and support structures. Wind turbines must be dismantled, and their foundations removed or stabilized. Battery systems must be safely disposed of or recycled due to the chemical materials they contain.

Communities hosting these projects often seek assurances that developers will be responsible for covering these future costs. Some jurisdictions require financial guarantees, such as decommissioning bonds or escrow accounts, to ensure that funds will be available when projects are retired.

Without such safeguards, there is a concern that long-term liabilities could fall on landowners or local governments if companies responsible for the projects no longer exist or are unable to fulfill their obligations decades later.

## Weighing the Full Picture

The hidden costs associated with the energy transition illustrate the complexity of shifting from one energy system to another. Renewable energy technologies offer important environmental advantages, particularly in reducing emissions during electricity generation. However, their development also involves financial investments, infrastructure expansion, resource extraction, and long-term management responsibilities.

Recognizing these factors allows for a more balanced understanding of the transition underway. Rather than viewing renewable energy as either a perfect solution or a flawed technology, it may be more productive to examine how policies, technologies, and local decision-making processes can evolve to address both the benefits and the challenges involved.

Energy systems have always involved trade-offs. The transition toward renewable energy represents the latest chapter in the ongoing effort to meet society's energy needs while minimizing environmental harm and ensuring long-term sustainability. Understanding the full range of costs and considerations is essential for making informed decisions about how that transition unfolds.

# Chapter 17
# Communities Push Back

As large-scale renewable energy projects spread into rural regions, a growing number of communities have begun to respond in ways that reflect both concern and engagement with the democratic process. What started in many places as confusion about unfamiliar development proposals has evolved into organized grassroots movements aimed at understanding, questioning, and influencing how these projects are approved.

For many residents, the issue is not simply about energy technology. It is about land use, local governance, and the future of communities that have long depended on agriculture and rural traditions. When solar farms, wind projects, or battery storage facilities are proposed on farmland, the decisions involved can affect landscapes, property values, and local economies for decades.

As a result, rural residents—farmers, homeowners, and local officials alike—are increasingly becoming active participants in the planning processes that shape their communities. Through public meetings, citizen groups, and policy advocacy, they are seeking to ensure that decisions about development are made thoughtfully and transparently.

## Citizen Activism and Community Organizing

One of the most visible responses to large renewable energy projects has been the rise of grassroots citizen groups. In many communities, these groups form organically when residents begin discussing development proposals and realize that others share similar questions or concerns.

Neighbors may begin by attending township meetings or planning commission hearings where projects are being discussed. Over time, these informal gatherings can grow into more organized efforts focused on research, education, and community engagement.

Some groups hold informational meetings to explain how renewable energy projects work and what potential impacts they may have on farmland, infrastructure, and local zoning laws. Others create online forums or newsletters to share updates about upcoming hearings and encourage residents to participate in the public comment process.

For many individuals, involvement in these activities represents their first experience with local government. People who had rarely attended township meetings before may suddenly find themselves studying zoning ordinances, reviewing project documents, and speaking at public hearings.

This form of grassroots engagement reflects a long tradition of civic participation in rural communities, where local decisions often depend heavily on citizen involvement.

## Legal Challenges and Policy Reform

As communities become more organized, some have turned to legal and policy strategies to ensure that development proposals are evaluated fairly and in accordance with existing regulations. Legal challenges may arise when residents believe that zoning rules have been misinterpreted, environmental reviews are incomplete, or public participation requirements have not been fully met.

In some cases, communities seek to strengthen local ordinances so they better address modern energy technologies. Many zoning regulations were written long before utility-scale solar farms or battery storage systems became common. Updating these ordinances can help clarify how such projects should be evaluated and what standards they must meet.

Local governments may revise setback requirements, establish safety guidelines, or define which zoning districts are appropriate for large energy facilities. Some communities temporarily pause approvals through moratoriums while they study potential impacts and rewrite regulations.

These policy changes reflect an effort to ensure that development occurs within a clear framework that protects both property rights and community interests.

## Protecting Local Decision-Making

A central theme in many grassroots movements is the desire to maintain local authority over land-use decisions. Townships and counties have historically held responsibility for zoning and planning within their boundaries. Residents often feel that local officials are best positioned to understand the unique characteristics of their communities.

When decisions about energy infrastructure are shifted to state or regional agencies, some residents worry that local voices may be overlooked. State-level policies aimed at accelerating renewable energy development can sometimes limit the ability of local governments to influence where projects are located.

For many rural communities, the debate is not solely about renewable energy itself but about who has the authority to decide how land is used. Residents often believe that the people who live closest to the land should have a meaningful role in determining its future.

Protecting local decision-making becomes a way of preserving both community identity and democratic participation.

## Searching for Balanced Solutions

Despite the conflicts that sometimes arise, many communities are not fundamentally opposed to renewable energy. Instead, they are searching for ways to balance the benefits of cleaner energy with the preservation of farmland, rural landscapes, and community stability.

Some communities explore alternative approaches to renewable energy development that reduce impacts on prime agricultural land. This may include encouraging solar installations on rooftops, industrial sites, or previously developed land rather than converting farmland.

Others focus on farmland preservation programs that protect highly productive soils from permanent development. By identifying areas where renewable energy projects are appropriate and areas where agriculture should remain the primary land use, communities attempt to guide development in ways that align with long-term planning goals.

These efforts reflect a recognition that energy systems must evolve while also acknowledging the importance of protecting land that sustains rural livelihoods.

## A Renewed Sense of Civic Engagement

The grassroots responses emerging across rural communities represent more than opposition to individual projects. They reflect a broader revival of civic participation in local government. Residents who once felt distant from policy decisions are now actively engaging in discussions about energy, land use, and the future of their communities.

Township halls that once hosted small gatherings may now be filled with residents eager to ask questions, express concerns, and participate in shaping local policy. While these meetings can sometimes be tense, they also demonstrate the resilience of local democracy.

Communities pushing back against large-scale development are often motivated by a desire to ensure that change happens thoughtfully and with respect for local values.

## A Defining Moment for Rural Communities

The debates surrounding renewable energy development have become defining moments for many rural regions. They challenge communities to consider how technological change should unfold in the landscapes where people live and work.

At the heart of these discussions lies a fundamental question: how can societies pursue cleaner energy while respecting the people and places most directly affected by the infrastructure required to produce it?

The answer will likely differ from one community to another. But the growing engagement of rural residents in these conversations ensures that the future of the countryside will not be shaped solely by distant policymakers or corporate investment.

Instead, it will also reflect the voices of the people who live on the land—and who believe that its future is worth defending.

# Chapter 18
# Communities Divided

One of the most difficult and often overlooked consequences of large-scale energy development in rural areas is the way it can divide communities that have long shared common histories and relationships. While debates about renewable energy are frequently framed in terms of technology, economics, or environmental policy, the most immediate effects are often felt at the local level—between neighbors, families, and friends who suddenly find themselves on different sides of the same issue.

In many rural communities, the arrival of solar farms, wind turbines, or battery storage facilities introduces choices that do not affect everyone equally. Some landowners may be offered financial opportunities through long-term lease agreements, while neighboring property owners may experience the impacts of the project without receiving similar benefits. This imbalance can create tensions that challenge the cohesion of communities that once worked together with little conflict.

Understanding these divisions requires looking closely at the perspectives of both those who support renewable energy projects and those who oppose them.

## Farmers Leasing Land

For many farmers, the decision to lease land for energy development is not taken lightly. Agriculture is an industry filled with uncertainty. Crop yields depend on weather conditions, commodity prices fluctuate unpredictably, and input costs for fuel, fertilizer, and equipment can rise sharply from year to year.

Against this backdrop of economic uncertainty, renewable energy developers sometimes offer lease agreements that provide stable income for decades. Solar or wind leases may guarantee annual payments that are not directly affected by market volatility or seasonal weather patterns.

For farmers facing financial pressure, these agreements can represent a form of security that traditional farming cannot always provide. Lease income may help families pay off debt, maintain ownership of land that might otherwise need to be sold, or support retirement planning for aging farmers.

In some cases, farmers view these agreements as a practical way to adapt to changing economic conditions while keeping their land within the family.

From this perspective, leasing land for renewable energy can appear to be a rational economic decision.

## Neighbors Opposing Projects

At the same time, neighboring residents may view the same projects very differently. Those who live near proposed energy installations may worry about how large solar arrays, wind turbines, or battery storage facilities could affect the character of the landscape they call home.

Concerns may include visual impacts, changes to farmland use, potential safety issues, or questions about property values. Residents who are not participating in lease agreements may feel that they are being asked to bear the impacts of development without receiving the financial benefits.

In communities where agriculture has long been the dominant land use, the transformation of fields into energy infrastructure can also raise concerns about the long-term future of farming. Some residents worry that once farmland begins to disappear, the rural character of the area may gradually change.

For those who see the countryside as something worth preserving, opposition to large-scale development can stem from a desire to protect landscapes that have remained relatively unchanged for generations.

## The Economic Divide

The difference between landowners who lease land and neighbors who oppose projects can create an economic divide within communities. Lease payments offered to participating landowners may be substantial, sometimes far exceeding the income that could be generated through traditional agricultural use.

Meanwhile, neighboring property owners may experience changes to their surroundings without any direct financial compensation.

This dynamic can lead to feelings of unfairness or resentment, particularly if projects occupy large areas of land. In some communities, residents worry that financial incentives offered by developers may encourage landowners to prioritize short-term economic gains over broader community interests.

These disagreements can place strain on relationships that were once built on shared experiences and mutual trust.

## Social and Community Impacts

The tensions created by energy development proposals can extend beyond economic concerns. Rural communities often rely on strong social bonds that have developed over many years. Families may have farmed side by side for generations, children may attend the same schools, and neighbors often assist one another during harvest seasons or difficult times.

When development proposals introduce competing interests, these relationships can become strained. Public meetings about energy projects sometimes become emotionally charged, as residents express deeply held views about the future of their community.

In some cases, neighbors who once worked together may stop speaking to one another, while community organizations may struggle to maintain unity amid disagreement.

These social impacts can be among the most difficult consequences of development debates because they affect the everyday relationships that define rural life.

## Searching for Common Ground

Despite these divisions, many communities continue searching for ways to find common ground. Some residents believe that renewable energy can be integrated into rural landscapes if projects are carefully planned and regulated. Others advocate for stronger zoning rules that protect farmland while allowing limited development in appropriate areas.

Community discussions may focus on issues such as setbacks, site selection, and long-term land restoration requirements. These conversations reflect efforts to balance economic opportunities with the desire to preserve rural character and agricultural land.

In many cases, the goal is not to eliminate disagreement but to ensure that decisions are made in ways that respect both individual property rights and community interests.

## A Test of Rural Resilience

The divisions created by renewable energy development illustrate the complex challenges facing rural communities as energy systems evolve. These debates test the ability of communities to navigate competing priorities while maintaining the relationships that sustain local life.

While disagreements may be inevitable, the process of discussing and addressing these issues can also strengthen civic engagement. Residents who participate in planning meetings, zoning hearings, and community forums become more involved in shaping the future of their towns.

In the end, the story of communities divided is not only about conflict. It is also about the resilience of rural communities as they confront difficult questions about land, energy, and the future of the places they call home.

# Chapter 19
# The Organized Voice

In every township across this country, there is a shared understanding—decisions that shape the land should come from the people who live on it. For generations, local meetings have served as the foundation of that principle. They are where neighbors gather, speak their minds, and help determine the future of their community.

But today, that process is changing.

Large-scale development projects, particularly in the energy sector, are no longer guided solely by engineering, economics, or necessity. Increasingly, they are accompanied by a different kind of infrastructure—one built not of steel or concrete, but of messaging, strategy, and influence.

Professional consulting firms now play a central role in these efforts. Their purpose is not to construct projects, but to construct support.

These firms begin by studying the community. They identify values, concerns, and potential points of resistance. From there, they build a communication strategy designed to resonate locally. Words like progress, opportunity, and responsibility are carefully chosen, not randomly spoken. They are tested, refined, and deployed with intent.

Next comes organization.

Field teams are assembled. Conversations begin—at doorsteps, on phones, and through digital platforms. Individuals who are open to the project are encouraged to participate. Over time, this creates a network of voices that can be activated when needed.

Then comes visibility.

Public meetings fill with attendees. Speakers arrive prepared. Messaging becomes consistent. What appears to be a spontaneous rise in community support often reflects a coordinated effort working quietly beneath the surface.

This does not mean that every voice in support is insincere. Many people speak honestly and believe in what they are saying. But the environment in which those voices emerge is no longer neutral. It is shaped, guided, and amplified.

And that distinction matters.

Because when decisions are made based on visible momentum alone, the true measure of community sentiment can become unclear. The loudest moment is not always the most representative one.

Local boards and planning commissions carry the weight of this reality. They must look beyond numbers, beyond appearances, and beyond pressure. Their responsibility is not to respond to organized energy, but to evaluate long-term impact.

- What will this mean for the land?
- What will this mean for water?
- What will this mean for the families who remain long after construction is complete?

These are the questions that matter.

For residents, awareness is the first step. Understanding how influence works does not weaken a community—it strengthens it. It allows individuals to recognize patterns, to think independently, and to ensure that their voices are grounded in truth rather than momentum.

The future of a township is not decided in a single meeting. It is shaped over time by the people who invest their lives in it.

And that future should never be determined by those who simply pass through, organize, and move on.

It belongs to those who stay.

# Chapter 20
# Food vs Energy

One of the most significant and emotionally charged questions surrounding large-scale renewable energy development is the growing tension between food production and energy production. As nations seek new ways to generate electricity while reducing reliance on fossil fuels, rural landscapes are increasingly being viewed as potential sites for energy infrastructure. At the same time, these same landscapes have long served as the foundation of agriculture, providing the food that sustains communities and populations around the world.

This intersection creates a fundamental question that many rural communities are now confronting: Should farmland produce food, energy, or both?

The answer is not simple. It involves economic pressures, environmental priorities, technological possibilities, and the long-term stewardship of land that must serve multiple purposes for future generations.

## The Importance of Agricultural Land

Farmland has always been one of the most essential resources in human society. The productivity of agricultural land determines how much food can be grown, how stable food supplies remain, and how rural economies function. In regions known for fertile soil and productive farming systems, agriculture often forms the backbone of the local economy.

Farmers depend on these lands to produce crops that feed both domestic and global markets. Corn, soybeans, wheat, vegetables, dairy products, and livestock all originate from fields and pastures managed by generations of farmers who have developed deep knowledge of the land.

Because productive farmland takes centuries to develop through natural soil formation processes, it cannot easily be replaced once it is converted to other uses. Urban expansion, highways, industrial facilities, and other forms of development have already reduced the total amount of farmland in many areas.

Renewable energy development introduces a new type of land use that competes with agriculture for the same space.

## The Land Requirements of Energy

Renewable energy systems, particularly solar and wind projects, require significant amounts of land to produce electricity on a large scale. Unlike traditional power plants that concentrate energy generation within relatively small facilities, renewable energy captures diffuse natural resources such as sunlight and wind.

To generate substantial amounts of electricity, developers must spread solar panels or wind turbines across large areas of land. A utility-scale solar installation, for example, may require hundreds or even thousands of acres depending on the amount of electricity it is designed to produce.

When these installations are built on farmland, the land is often removed from traditional crop production for decades. Even if vegetation remains beneath solar panels, the land may no longer function as part of an active farming operation.

This shift transforms land that once produced food into land that produces electricity.

## Economic Incentives

The economic realities facing farmers also play an important role in the food-versus-energy debate. Agriculture can be a challenging industry with fluctuating commodity prices, unpredictable weather, and rising operating costs.

Renewable energy developers sometimes offer long-term lease agreements that provide stable annual payments for the use of farmland. These payments can exceed what farmers might earn from crop production on the same acreage.

For some landowners, these leases provide financial stability that helps keep the farm in the family. For others, they may represent an opportunity to diversify income while continuing agricultural activities on other parts of their land.

However, these economic incentives can also shift how land is valued. Fields that once held value primarily for their agricultural productivity may begin to be evaluated based on their potential for energy generation.

## Global Food Security

Beyond local economic considerations lies a broader question about global food production. As the world's population continues to grow, demand for food is expected to increase significantly. Agricultural land will play a critical role in meeting that demand.

If large areas of farmland are converted to energy production, some analysts worry that food production capacity could eventually be affected. While technological advances in agriculture have improved yields, the availability of productive land remains a fundamental constraint.

This concern has led some researchers and policymakers to explore alternative approaches that allow renewable energy development without reducing farmland availability.

## Dual-Use Approaches

One emerging concept aimed at balancing these competing demands is known as agrivoltaics. In these systems, solar panels are installed in ways that allow crops or livestock to continue using the land beneath or between panel rows.

For example, sheep grazing beneath solar arrays has been implemented in some regions, while certain crops that tolerate partial shade may be grown under elevated panel structures.

Although these approaches show promise in some situations, they are not universally applicable. Many crops require full sunlight and the ability to use large agricultural machinery, which may not be compatible with solar panel layouts.

As a result, agrivoltaic systems may work best in specific agricultural contexts rather than serving as a universal solution.

## The Long-Term Perspective

At its core, the food-versus-energy debate is about long-term land stewardship. The decisions communities make today about how farmland is used will shape landscapes for decades to come.

Energy infrastructure can remain in place for thirty years or more, meaning that fields converted today may not return to traditional farming until well into the future. This timeline raises questions about what future generations will inherit.

Will the countryside continue to function primarily as a landscape of food production? Or will it increasingly become a landscape where agriculture shares space with energy infrastructure?

## A Question of Balance

Ultimately, the challenge facing many communities is not choosing between food and energy entirely, but determining how the two can coexist in a way that protects both. Renewable energy may play an important role in addressing environmental concerns, but farmland remains essential for feeding populations and sustaining rural economies.

Finding the right balance will require thoughtful planning, careful land-use policies, and open dialogue among farmers, residents, policymakers, and energy developers.

The fields that stretch across rural landscapes today represent more than open land. They represent the intersection of two critical needs—food and energy—both of which will shape the future of society.

How those fields are used in the decades ahead may ultimately determine whether the countryside remains primarily a source of food, becomes a source of energy, or evolves into a landscape that must serve both roles simultaneously.

# Chapter 21
# Protecting Farmland

Across rural landscapes, farmland represents far more than open space. It is the foundation of food production, the backbone of rural economies, and a living system that supports soil health, water cycles, and local ecosystems. As development pressures increase—from housing growth to industrial infrastructure and renewable energy projects—communities are increasingly asking how farmland can be protected for future generations.

The question of protecting farmland is not simply about preventing development. Instead, it is about ensuring that agricultural land remains available and productive while still allowing societies to meet evolving needs for energy, housing, and economic growth.

In recent years, the rise of large-scale renewable energy projects has added a new dimension to this conversation. Communities must now consider how to preserve farmland while also participating in the transition toward new energy systems.

## Farmland Preservation Policies

Many regions have adopted farmland preservation policies designed to protect agricultural land from permanent conversion to other uses. These policies recognize that once farmland is developed or industrialized, restoring it to productive agriculture can be extremely difficult.

Common farmland preservation strategies include:

- Agricultural zoning districts that limit non-farm development in rural areas
- Purchase of development rights (PDR) programs that compensate landowners for agreeing to keep land in agriculture

- Conservation easements that legally restrict certain types of development on farmland
- Farmland protection plans are incorporated into local or regional master planning documents

These policies are designed to ensure that large blocks of agricultural land remain intact rather than gradually being fragmented by development.

In many farming regions, preserving farmland also helps maintain the infrastructure necessary for agriculture, such as grain elevators, equipment suppliers, and transportation networks.

## Agricultural Priorities

Beyond formal policies, protecting farmland often reflects a broader commitment to agriculture as a priority land use. Rural communities that depend on farming frequently view agriculture not only as an industry but also as a defining cultural and economic feature of their region.

Agricultural priorities may include:

- Maintaining productive soils
- Supporting family farms
- Preserving agricultural economies
- Ensuring long-term food production

Local planning documents sometimes designate agriculture as the preferred or primary land use in certain areas. This designation helps guide zoning decisions and development proposals.

When renewable energy projects are proposed on agricultural land, local governments must decide whether those projects align with the community's stated priorities.

In some cases, communities may determine that certain types of development are compatible with agriculture. In others, they may decide that preserving farmland should take precedence.

## The Role of Soil and Land Quality

Not all farmland is equal. Some regions contain prime agricultural soils that are particularly valuable for crop production due to their fertility, drainage, and structure. These soils often developed over thousands of years through natural processes.

Because prime farmland is relatively rare and highly productive, many farmland preservation efforts focus on protecting these areas first. Losing highly productive soils to permanent development can have long-term consequences for regional food production.

For this reason, some communities adopt land-use policies that steer industrial development toward less productive land while preserving areas with the highest agricultural value.

## Balancing Energy and Food Production

The challenge facing many rural communities today is how to balance farmland preservation with the growing demand for renewable energy infrastructure. Solar farms, wind projects, and battery storage facilities require space, and developers often seek locations with large open areas and access to transmission lines.

Unfortunately, these characteristics frequently overlap with agricultural regions.

Communities are therefore exploring strategies that allow renewable energy development while minimizing impacts on farmland. Possible approaches include:

- Directing projects toward previously developed or industrial land
- Prioritizing renewable energy on brownfields or unused sites
- Limiting projects on prime agricultural soils
- Encouraging dual-use systems where agriculture and energy coexist

For example, some solar projects incorporate grazing livestock beneath panel arrays or plant pollinator-friendly vegetation that benefits surrounding crops. These approaches attempt to maintain some agricultural function while generating electricity.

However, such systems are not suitable for all crops or farming practices, and their long-term effectiveness remains an area of ongoing study.

## Stewardship and Future Generations

Protecting farmland ultimately reflects a philosophy of long-term stewardship. Farmers and rural communities often think in terms of generations rather than short-term economic cycles. Decisions about land use today will shape what future farmers inherit.

Fields that produce crops today may continue feeding communities decades from now if the land is carefully managed and preserved. Once farmland is permanently converted, however, that opportunity may be lost.

For many rural residents, farmland protection is therefore about safeguarding a resource that belongs not only to the present generation but also to those who will follow.

## A Landscape of Responsibility

As societies navigate the transition toward new energy systems, the question of protecting farmland will remain central to discussions about rural land use. Renewable energy may offer environmental benefits and economic opportunities, but agriculture remains essential for sustaining populations and rural economies.

Balancing these needs requires thoughtful planning, clear policies, and open dialogue between farmers, policymakers, developers, and communities.

The countryside is more than a place where infrastructure can be built. It is a living landscape shaped by soil, water, crops, and human stewardship. Protecting farmland means recognizing the responsibility that comes with managing that landscape—ensuring that it continues to provide food, livelihoods, and environmental benefits for generations to come.

# Chapter 22
# The Future of Rural Land

The debate over renewable energy development ultimately leads to a larger and more enduring question: what will the future of rural land look like? As energy systems evolve and societies work to address environmental challenges, rural landscapes are increasingly being asked to serve new purposes beyond traditional agriculture. Solar arrays, wind turbines, battery storage facilities, and transmission corridors are becoming part of the countryside in ways that few communities anticipated only a generation ago.

At the same time, farmland continues to play an essential role in food production, rural economies, and environmental stewardship. Fields that produce crops today may have been cultivated by generations of farmers, forming part of a long tradition of land management and agricultural knowledge.

The challenge facing many communities today is how to balance these two realities: the growing demand for renewable energy and the enduring importance of preserving farmland.

## A Landscape in Transition

Rural landscapes have always evolved over time. Agricultural technology has transformed the way crops are planted and harvested, irrigation systems have reshaped water management, and new transportation networks have connected farms to global markets. These changes reflect the ability of rural communities to adapt to shifting economic and technological conditions.

The expansion of renewable energy represents another phase in this long history of change. As governments and utilities seek to reduce reliance on fossil fuels, rural land is increasingly viewed as a key component of the energy transition. Large open areas that once supported only crops or livestock are now being considered as potential sites for electricity generation.

For some communities, this transformation represents economic opportunity and participation in new industries. For others, it raises concerns about the long-term preservation of agricultural landscapes.

## Competing Demands for Land

One of the central challenges shaping the future of rural land is the competition between different land uses. Farmland must now accommodate a growing range of needs, including food production, housing development, conservation efforts, and energy infrastructure.

Renewable energy technologies often require large areas of land to generate electricity at scale. Solar farms may occupy hundreds or thousands of acres, while wind projects spread turbines across wide landscapes. Transmission lines connecting these facilities to population centers can extend for many miles.

At the same time, global demand for food continues to increase as populations grow. Maintaining productive farmland will remain essential for meeting these needs. This creates a complex balancing act between the desire to expand renewable energy and the need to preserve agricultural capacity.

Communities facing these choices must evaluate how land can serve multiple purposes without sacrificing the resources that sustain rural life.

## Exploring Balanced Solutions

In response to these challenges, some communities and policymakers are exploring ways to balance energy development with farmland preservation. One approach involves directing renewable energy projects toward locations that have already been developed or disturbed, such as industrial sites, brownfields, abandoned mines, or large commercial rooftops.

Another strategy focuses on dual-use systems that allow certain forms of agriculture to coexist with renewable energy installations. For example, some solar projects incorporate grazing areas for livestock or plant pollinator-friendly vegetation beneath panel arrays. While these approaches may not fully replace traditional farming, they represent efforts to integrate energy production with agricultural landscapes.

Land-use planning can also play an important role in identifying which areas are most suitable for renewable development and which should remain dedicated to agriculture. By establishing clear guidelines and zoning policies, communities can guide development in ways that reflect local priorities.

## Stewardship and Long-Term Thinking

At the heart of the debate about rural land lies the concept of stewardship. Farmers often view their land not only as an economic asset but also as a responsibility to future generations. Soil health, water quality, and sustainable farming practices are all part of maintaining land that will remain productive for decades to come.

Energy infrastructure introduces new considerations into this tradition of stewardship. Decisions about land use today will shape landscapes for many years, particularly when energy facilities operate for several decades before decommissioning.

Communities must therefore consider how these projects fit into long-term visions for the countryside. Will renewable energy become a permanent feature of rural landscapes? Or will future technologies allow energy production to shift toward different locations or methods?

The answers to these questions remain uncertain, but they will influence how rural regions evolve.

## The Role of Local Communities

Perhaps the most important factor shaping the future of rural land is the role of the communities that live there. Local residents, farmers, and elected officials are often the people most directly affected by decisions about land use. Their perspectives help determine whether renewable energy projects are welcomed, modified, or opposed.

Public participation in planning processes allows communities to weigh the benefits of new technologies against the potential impacts on farmland and rural character. While these discussions can sometimes be contentious, they reflect the importance of local voices in shaping land-use decisions.

When communities actively engage in these conversations, they help ensure that the future of rural land is not determined solely by external economic or political forces.

## A Landscape Shaped by Choices

The future of rural land will not be determined by technology alone. It will be shaped by the choices made by farmers, policymakers, investors, and communities. Decisions about where renewable energy projects are located, how farmland is protected, and how infrastructure is built will all influence the character of the countryside.

In many ways, rural landscapes have always reflected the priorities of the societies that depend on them. The barns, fields, and roads that define agricultural regions today are the result of generations of decisions about how land should be used.

The energy transition introduces a new set of choices into that long history. Whether rural land becomes a primary source of food, energy, or a combination of both will depend on how communities balance these competing needs.

The fields that stretch across the countryside today are not only places where crops grow—they are landscapes of possibility. How they are used in the years ahead will shape not only the future of rural communities but also the broader relationship between land, energy, and society itself.

# Chapter 23
# The Farmland Question

For centuries, farmland has served as one of the most fundamental resources in human civilization. It is the land that feeds populations, supports rural economies, and sustains generations of farming families. Fields are more than open spaces—they are living systems that produce crops, nurture soil ecosystems, and anchor communities whose livelihoods depend on the rhythms of agriculture.

As renewable energy projects expand across rural regions, however, an increasingly important question has emerged: what happens when farmland becomes energy infrastructure?

The conversion of agricultural land into solar installations, wind facilities, and associated electrical infrastructure introduces a long-term shift in how land is used. While renewable energy projects may contribute to environmental goals such as reducing carbon emissions, they also compete with agriculture for one of society's most valuable resources—productive land.

The farmland question is therefore not simply about energy technology. It is about how societies balance competing needs for food, energy, and land stewardship.

## A Finite Resource

One of the most important realities about farmland is that it is limited. High-quality agricultural soils develop over long periods of time through natural processes involving weathering, biological activity, and organic matter accumulation. In many regions, the most productive soils took thousands of years to form.

Once farmland is permanently converted to other uses—such as urban development, highways, or industrial infrastructure—it is rarely restored to its original productivity. While some land can be reclaimed or rehabilitated, the process often requires significant time and investment, and the results may not fully match the original soil quality.

Renewable energy infrastructure introduces a new form of competition for this limited land base. Solar installations may occupy hundreds or even thousands of acres for decades, effectively removing that land from agricultural production for the duration of the project.

This raises important questions about how much farmland societies can afford to convert without affecting long-term food production capacity.

## Agriculture and the Rural Economy

Farmland supports far more than crop production alone. Rural economies are often built around agricultural systems that include equipment suppliers, grain elevators, livestock operations, food processing facilities, and transportation networks.

When farmland disappears, these interconnected systems can also be affected. Fewer acres of crops may mean less business for local agricultural service providers. Communities that rely heavily on farming may experience gradual changes in employment patterns and economic activity.

In regions where agriculture has shaped local identity for generations, the loss of farmland can also represent a cultural shift. Farms often serve as family legacies passed down through multiple generations, connecting people to both the land and their community's history.

When agricultural landscapes change, the cultural and economic fabric of rural life can change as well.

## Long-Term Land Commitments

Renewable energy projects are typically designed to operate for twenty-five to thirty years, and sometimes longer. During that time, solar panels, electrical infrastructure, and maintenance roads remain in place across the land.

Although the land beneath these installations may still support vegetation or limited agricultural activities such as grazing, it no longer functions as traditional farmland. Crops that require large machinery or regular cultivation cannot easily coexist with rows of solar panels or other infrastructure.

The long-term nature of these projects means that decisions made today can shape land use patterns for decades. Communities considering energy development proposals must therefore evaluate not only immediate economic benefits but also the long-term implications for agriculture and land stewardship.

## Food Production and Global Demand

Another dimension of the farmland question involves global food security. As the world's population continues to grow, demand for food is expected to increase significantly. Maintaining productive agricultural land will play a critical role in meeting that demand.

Some analysts argue that renewable energy development should prioritize land that has already been disturbed or developed—such as industrial sites, parking areas, rooftops, or brownfields—rather than converting prime farmland. Others suggest that certain forms of dual-use agriculture, sometimes called agrivoltaics, may allow crops or livestock to coexist with solar installations under certain conditions.

However, the practicality and scalability of these approaches remain topics of ongoing debate. Not all agricultural systems are compatible with solar infrastructure, and the economics of combining energy production with farming activities can vary widely depending on location and crop type.

## Stewardship and Long-Term Thinking

At its core, the farmland question is about stewardship. Farmers often view themselves as caretakers of the land, responsible for maintaining soil health and productivity for future generations. Decisions about land use, therefore, carry both economic and ethical implications.

Communities facing proposals for large energy projects must weigh a variety of factors: the need for cleaner energy, the economic opportunities offered by renewable development, and the long-term importance of preserving farmland.

These decisions rarely have simple answers. Each community may arrive at different conclusions based on its unique circumstances, economic priorities, and values.

What remains clear, however, is that farmland represents a resource that extends far beyond any single generation. Once converted, its role in agriculture may be difficult to reclaim.

## A Question That Will Shape the Future

As renewable energy continues to expand, the farmland question will likely become one of the defining debates of the energy transition. The choices societies make about how land is used will influence not only the structure of future energy systems but also the ability to sustain agricultural production and rural communities.

The fields that stretch across rural landscapes today represent centuries of cultivation and stewardship. Deciding how those fields should be used in the decades ahead is one of the most important land-use decisions facing modern societies.

In the end, the farmland question asks something deeper than where energy projects should be built. It asks how we value land itself—and what responsibilities we carry in deciding how it will be used for generations to come.

# Chapter 24
# The Line in the Soil

Across rural landscapes, debates over renewable energy development often appear to focus on infrastructure—solar panels, wind turbines, battery storage systems, and transmission lines. Yet beneath these discussions lies a deeper issue that cannot be measured in megawatts or acres alone. For many rural communities, the debate is about something far more fundamental: the protection of land, livelihood, and a way of life that has defined the countryside for generations.

The phrase "a line in the soil" represents a moment when communities decide that certain values must be defended. It is not necessarily a rejection of change or progress. Rather, it is a recognition that some aspects of rural life—farmland, community identity, and local self-determination—are worth preserving even as new technologies reshape the world.

Understanding what rural communities are fighting to protect requires looking beyond the technical details of energy development and examining the deeper connections between people and the land.

## Land as Legacy

For farmers and rural families, land is rarely viewed simply as property or a financial asset. It often represents a legacy passed from one generation to the next. Farms that exist today may have been cultivated by grandparents, great-grandparents, or earlier settlers who cleared fields, built barns, and developed the soil through decades of hard work.

Each season of planting and harvest becomes part of a long chain of stewardship. Farmers learn to understand the particular characteristics of their land—how water moves through fields, where soil is richest, and how weather patterns affect crops.

Because of this deep connection, the land itself becomes part of family history. Decisions about how it will be used carry emotional and cultural weight far beyond economic considerations.

When rural residents draw a line in the soil, they are often defending this legacy.

## Protecting Farmland

One of the central concerns voiced by rural communities is the long-term preservation of farmland. Productive agricultural land is a finite resource. Once it is converted into other forms of development, restoring it to full agricultural use can be difficult.

For generations, farmers have worked to maintain soil health, improve drainage systems, and develop sustainable farming practices that ensure land remains productive over time. These efforts reflect a commitment to stewardship that extends beyond individual ownership.

When large infrastructure projects are proposed on farmland, some residents worry that these lands may gradually disappear from agricultural use. Even if land remains technically undeveloped beneath solar panels or other installations, the shift away from active farming can change how the land functions within the broader agricultural system.

Protecting farmland, therefore, becomes a way of protecting food production, local economies, and the long-term sustainability of rural regions.

## Preserving Rural Character

Rural landscapes possess a character shaped by open fields, barns, tree lines, and quiet roads stretching between farms. These landscapes are not only visually distinctive but also culturally meaningful. They reflect patterns of life built around agriculture, seasonal work, and close-knit communities.

For many residents, preserving rural character means maintaining the balance between development and open land that defines the countryside. Large industrial projects—regardless of their purpose—can alter this balance by introducing infrastructure that changes how the landscape looks and functions.

When communities speak about protecting rural character, they are often expressing a desire to ensure that future generations will inherit landscapes that still resemble the places where they grew up.

## Defending Local Voice

Another important dimension of the line in the soil involves the role of local communities in shaping their own future. Rural residents often believe that those who live closest to the land should have a meaningful voice in decisions about how it is used.

Township boards, planning commissions, and zoning hearings provide forums where citizens can participate directly in shaping land-use policies. When communities organize to attend meetings, speak during public comment periods, and engage with local officials, they are exercising a form of grassroots democracy.

Drawing a line in the soil can therefore represent a defense of local decision-making as much as a defense of the land itself.

## A Question of Values

The debates surrounding energy development often bring competing values into focus. Some residents emphasize the importance of transitioning to cleaner energy systems and reducing environmental impacts associated with fossil fuels. Others focus on protecting farmland, preserving rural landscapes, and maintaining community control over land use.

These perspectives are not always mutually exclusive, but balancing them can be difficult. Communities must decide how much change they are willing to accept and what aspects of their landscape they believe should remain unchanged.

The line in the soil represents the point at which communities begin defining those boundaries.

## A Moment of Choice

Ultimately, the line in the soil is not about resisting progress. It is about recognizing that the land carries meaning that extends beyond economic or technological considerations. Fields are not simply empty spaces waiting for development—they are places where food is grown, families work together, and traditions continue.

When rural communities stand up to protect their land, they are participating in a long tradition of stewardship that values both the productivity of the soil and the identity of the people who depend on it.

The decisions made during these debates will shape the countryside for generations to come. Whether the future landscape includes farms, energy infrastructure, or a combination of both, the voices of rural communities will continue to play a crucial role in determining how that future unfolds.

In the end, the line in the soil is a reminder that land is not only a resource. It is a foundation for community, heritage, and responsibility to those who will inherit it next.

## The Line in the Soil

Every community eventually reaches a moment when it must decide what it stands for.

For rural townships facing large energy developments, that moment often arrives in the form of a vote. A planning commission meeting stretches late into the evening. Residents line the walls of a township hall that suddenly feels too small for the number of people who have come to speak.

Farmers who rarely attend public meetings step to the microphone.

Young families speak about the future they hope to build.

Neighbors who have known each other for decades sometimes find themselves standing on opposite sides of the same issue.

The debate is rarely simple.

Renewable energy promises progress and environmental benefits. But the land beneath those projects carries its own history and value—one that cannot easily be replaced once it is gone.

For many rural communities, the question is not whether energy should evolve. The world will always need new sources of power and new technologies.

The real question is whether those decisions will respect the people and places most directly affected.

In the end, every township draws its own line.

Sometimes that line is drawn in zoning ordinances, sometimes in courtrooms, and sometimes simply in the determination of residents who refuse to let decisions about their land be made without their voice.

But wherever that line is drawn, it represents something deeper than a policy dispute.

It represents a community deciding that the ground beneath their feet is worth defending.

# Chapter 25
# What it Comes Down To

## Safety, Environment, and Infrastructure

As renewable energy technologies expand across modern power systems, attention is increasingly turning toward the broader questions of safety, environmental impacts, and the infrastructure required to support these new forms of energy production. Solar panels and wind turbines often symbolize clean energy in public messaging, but the systems that support them—particularly large-scale battery storage and expanded electrical networks—introduce new technical and environmental considerations.

While renewable energy technologies can reduce emissions during electricity generation, they also bring challenges that must be addressed carefully to ensure that the transition to cleaner energy remains safe, reliable, and sustainable over the long term.

Understanding these issues requires examining three interconnected areas: battery safety, electrical grid reliability, and the environmental footprint of large-scale energy infrastructure.

## Battery Storage and Safety

One of the most significant technological developments accompanying renewable energy growth is the rapid expansion of battery energy storage systems. These systems are designed to store electricity when renewable generation is high and release it later when demand increases or when sunlight and wind are unavailable.

Large battery installations typically use lithium-ion technology similar to the batteries found in smartphones and electric vehicles, but at a much larger scale. These grid-scale battery systems are often housed in large container units arranged across industrial sites.

While battery storage offers important benefits for balancing renewable energy supply, it also introduces safety considerations. Under certain conditions, lithium-ion batteries can experience a phenomenon known as thermal runaway, where overheating in one battery cell triggers a chain reaction that spreads to neighboring cells. This process can lead to fires that burn at very high temperatures and may be difficult to extinguish.

Emergency response teams and fire departments in some communities have begun training specifically to handle incidents involving battery storage facilities. Developers and engineers are also working to improve safety through monitoring systems, fire suppression technology, and facility design that limits the spread of potential fires.

As battery storage becomes more widespread, ensuring that safety standards and emergency response protocols keep pace with the technology remains an important priority.

## Grid Reliability in a Changing Energy System

Another critical issue involves the reliability of the electrical grid as renewable energy sources become more prominent. Traditional power plants fueled by coal, natural gas, or nuclear energy typically generate electricity continuously and can adjust output in response to changing demand.

Solar and wind power operate differently. Their output depends on natural conditions such as sunlight and wind speed, which can fluctuate throughout the day or change rapidly due to weather patterns. This variability introduces new challenges for grid operators responsible for maintaining a constant balance between electricity supply and demand.

Battery storage systems help address some of these fluctuations, but they cannot yet provide long-duration energy storage on the same scale as conventional fuel sources. As a result, grid operators often rely on a combination of technologies—including backup generation, expanded transmission networks, and advanced forecasting tools—to maintain stability.

Modern power grids are therefore becoming more complex systems that must integrate multiple energy sources, storage technologies, and digital monitoring systems to ensure reliable electricity delivery.

## Infrastructure Expansion

The growth of renewable energy also requires substantial expansion of physical infrastructure. Solar farms, wind projects, and battery storage facilities must all connect to the electrical grid through substations, transformers, and transmission lines capable of carrying electricity across long distances.

In many cases, renewable energy projects are located in rural areas where land is available and natural resources are favorable. However, the electricity they produce is often intended for urban centers where demand is highest. This geographic separation means that new transmission lines must be built to transport power from generation sites to population centers.

Constructing transmission infrastructure can involve clearing land corridors, erecting large steel towers, and installing miles of high-voltage cables. While these projects are essential for integrating renewable energy into the grid, they can also raise concerns among communities about landscape impacts and land use.

Infrastructure planning, therefore, plays a critical role in determining how renewable energy development affects both rural and urban environments.

## Environmental Considerations

Although renewable energy technologies produce electricity without combustion emissions, they still have environmental footprints associated with their construction, operation, and eventual disposal. Solar panels and wind turbines require raw materials such as metals, rare earth elements, and specialized manufacturing processes. Battery systems rely on minerals such as lithium, cobalt, and nickel.

In addition to material extraction, the installation of large energy facilities can affect local ecosystems. Construction may disturb wildlife habitats, alter drainage patterns, or change land use over large areas.

Efforts are increasingly underway to minimize these impacts through improved site selection, habitat protection measures, and recycling programs designed to recover materials from retired energy equipment. However, as renewable infrastructure continues to expand, careful environmental planning will remain essential.

## Balancing Innovation and Responsibility

The transition toward renewable energy represents one of the most significant transformations of modern infrastructure. New technologies promise cleaner electricity and reduced reliance on fossil fuels, but they also introduce new systems that must be designed and managed responsibly.

Safety protocols, grid reliability strategies, and environmental safeguards are all essential components of this transition. Engineers, policymakers, and communities must work together to ensure that renewable energy infrastructure not only meets energy needs but also protects people, landscapes, and ecosystems.

As renewable technologies continue to evolve, the conversation about energy will increasingly involve not only how electricity is generated but also how the systems supporting that generation are designed, regulated, and maintained for the long term.

# Bibliography


**Government & Policy Sources**

U.S. Department of Energy. (2023). Energy Storage and Grid Modernization Reports. Washington, DC.

U.S. Energy Information Administration (EIA). (2024). Annual Energy Outlook. Washington, DC.

U.S. Environmental Protection Agency (EPA). (2023). Renewable Energy and Land Use Guidelines. Washington, DC.

U.S. Department of Agriculture (USDA). (2022). Natural Resources Conservation Service: Soil Health and Farmland Preservation. Washington, DC.

Michigan Legislature. (2006). Michigan Zoning Enabling Act (Public Act 110 of 2006). Lansing, MI.

Michigan Public Service Commission (MPSC). (2023). Energy Infrastructure Siting and Regulatory Framework. Lansing, MI.

**Energy & Technology Sources**

International Energy Agency (IEA). (2023). Renewables 2023: Analysis and Forecast to 2028. Paris, France.

National Renewable Energy Laboratory (NREL). (2023). Utility-Scale Solar and Energy Storage Overview. Golden, CO.

U.S. Department of Energy. (2022). Grid Energy Storage Technology Cost and Performance Assessment. Washington, DC.

Tesla, Inc. (2023). Megapack Technical Overview. Palo Alto, CA.

Canadian Solar. (2023). SolBank Battery Energy Storage System Specifications.

**Agriculture, Soil & Land Use**

American Farmland Trust. (2023). Farms Under Threat: The State of America's Farmland. Washington, DC.

U.S. Department of Agriculture (USDA). (2022). Soil Health and Conservation Practices. Washington, DC.

Food and Agriculture Organization of the United Nations (FAO). (2022). The Future of Food and Agriculture. Rome, Italy.

Midwest Regional Planning Commission. (2021). Agricultural Land Preservation Strategies.

**Environmental & Water Resources**

U.S. Geological Survey (USGS). (2023). Groundwater and Aquifer Systems in the United States. Reston, VA.

Environmental Protection Agency (EPA). (2022). Stormwater Management and Runoff Control Guidelines. Washington, DC.

National Oceanic and Atmospheric Administration (NOAA). (2023). Climate and Weather Impacts on Infrastructure. Washington, DC.

**Local Governance & Land Use**

American Planning Association (APA). (2022). Planning for Renewable Energy. Chicago, IL.

Michigan Townships Association. (2023). Zoning, Special Land Uses, and Local Authority Guide. Lansing, MI.

Lincoln Institute of Land Policy. (2021). Land Use, Zoning, and Property Rights in the United States. Cambridge, MA.

**Community, Policy, and Economic Perspectives**

Brookings Institution. (2022). The Economics of Renewable Energy Development. Washington, DC.

Heritage Foundation. (2021). Energy Policy and Economic Impacts. Washington, DC.

World Bank. (2023). Energy Transition and Land Use Considerations. Washington, DC.

In addition to the sources listed above, this book incorporates local township documents, planning commission minutes, public records, community surveys, and firsthand observations from rural residents and farmers directly impacted by energy development. These lived experiences provide essential context for understanding how energy development is shaping communities at the local level.

www.ingramcontent.com/pod-product-compliance
Lightning Source LLC
LaVergne TN
LVHW050921210926
848482LV00012B/490

* 9 7 9 8 9 0 1 9 0 6 1 2 5 *